深入理解 React Router
从原理到实践

李杨韬 / 著

电子工业出版社
Publishing House of Electronics Industry
北京·BEIJING

内 容 简 介

本书从基础内容出发，详细梳理了浏览器的基础导航能力、history 库的使用及原理，以及 React Hooks 等基础知识，帮助读者学习并掌握 React Router 的前驱知识。同时，本书从 React Router 的发展历程、技术演变出发，介绍了 React Router 的设计思路，并通过路由器、路由端口、导航三要素，引出 React Router 的基本使用方法，并对源码进行了全面解析。此外，本书各章使用 React Hooks 穿插了 30 余个案例，在兼顾实用性、可扩展性的同时，为读者学习与理解 React Router 提供了全面丰富的素材。通过阅读本书，读者不仅可以融会贯通地掌握 React Router，而且能提升对前端路由的认识，并掌握前端领域路由的设计思路与方法。

本书适合有 JavaScript 或 TypeScript 基础，想要学习使用 React Router，或对 React Router 实现原理和工程实践感兴趣的开发者阅读。

未经许可，不得以任何方式复制或抄袭本书之部分或全部内容。
版权所有，侵权必究。

图书在版编目（CIP）数据

深入理解 React Router：从原理到实践 / 李杨韬著. —北京：电子工业出版社，2021.3
ISBN 978-7-121-40608-9

Ⅰ. ①深… Ⅱ. ①李… Ⅲ. ①移动终端－应用程序－程序设计 Ⅳ. ①TN929.53

中国版本图书馆 CIP 数据核字（2021）第 035987 号

责任编辑：宋亚东　　　　　　特约编辑：田学清
印　　刷：三河市华成印务有限公司
装　　订：三河市华成印务有限公司
出版发行：电子工业出版社
　　　　　北京市海淀区万寿路 173 信箱　　邮编：100036
开　　本：787×980　1/16　印张：22.25　字数：498 千字
版　　次：2021 年 3 月第 1 版
印　　次：2021 年 3 月第 1 次印刷
定　　价：89.00 元

凡所购买电子工业出版社图书有缺损问题，请向购买书店调换。若书店售缺，请与本社发行部联系，联系及邮购电话：（010）88254888，88258888。
质量投诉请发邮件至 zlts@phei.com.cn，盗版侵权举报请发邮件至 dbqq@phei.com.cn。
本书咨询联系方式：（010）51260888-819，faq@phei.com.cn。

序
Foreword

在我刚进入前端开发行业的时候，业界普遍采用多页面应用的模式，前端页面依附于服务端应用平台并作为展示模板，由服务端响应页面间的跳转请求。随着前端技术的飞速发展，AJAX逐渐成为流行的前后端通信技术，使得页面不再需要在页面加载或表单提交时同步地向服务端提交或获取数据，而是异步地向服务端提交或获取数据。从此，前后端分离时代来临。

前后端分离时代的来临，使得前后端分工逐渐趋于独立，前端页面不再依附于服务端，页面直接部署于 CDN 中。页面元素通过组件搭建而成，并直接在前端进行状态管理。原有的多页面应用也将页面视为组件，开始出现单页面应用（Single Page Application，SPA）。

优秀的单页面应用为用户提供了近似本地软件的体验，早期最有名的便是 Gmail。Gmail 率先为业界展示了单页面应用的魅力，用户在使用 Gmail 时，与使用本地邮件客户端并无差别，用户无须等待页面刷新。

单页面应用要实现加载后无刷新的体验，除了需要采用 AJAX 来代替表单提交，还离不开前端路由器。在浏览器环境当中，用户会对页面进行后退、前进、保存书签、分享 URL 等操作，这些操作原本是浏览器向服务端发送请求，由服务端路由进行响应支持的。而在前后端分离、单页面的架构下，对用户此类行为的处理自然需要交到前端，于是业界很快开始抽象出前端路由。

React 作为当前业界流行的三大前端 UI 框架之一，其配套的前端路由 React Router 使用率也极高。截至 2019 年年末，react-router 包每周下载量超过 250 万次。前端工程师很有必要了解路由的使用方式和运作原理，如果使用的是 React，则很有必要了解 React Router。

本书详细地介绍了 React Router 的历史演变、使用方式及运作原理。

在使用方式上，书中提供了基础场景的示例，读者可以学到 React Router 的使用方式。在后面的进阶实战示例章节中，读者也可以学到在工程实践中如何运用前端路由。全书还包括对 React 当前最新的 React Hooks API 的讲解。

在运作原理相关的章节中，从最基础的 URL、浏览器 history API，到 React Router 的源码实现，都有详细的讲解，这对希望了解前端路由实现原理的工程师来说会有不少的帮助。

本书基本囊括了所有与 React Router 相关的技术知识和细节，推荐想深入了解现代前端路由的工程师阅读。

叶俊星（Jasin Yip）

全栈工程师，先后就职于美团网、阿里巴巴

前 言
Preface

React Router 自 2014 年 2 月发布第一个版本，到 2019 年 10 月发布 v5.1.2，已经经历了大大小小 130 多个版本的迭代演进。从 2020 年 1 月的 GitHub 数据来看，有超过 80 万个仓库的依赖包含了 React Router，同时 NPM 上依赖 React Router 的 react-router 包数量已超过 5200 个。2019年，在 NPM 源中，React Router 的 react-router 包周平均下载量超过 250 万次。以上数据足以说明，React Router 几乎成了 React 生态中路由的标准解决方案。

鉴于目前相关系统介绍 React Router 的技术图书比较匮乏，特编写此书以飨读者。本书基于 React 开发框架，在提供基础知识的同时，将基于 React Router v5.1.2 进行介绍。以下是各章节内容。

第 1 章，主要介绍与导航相关的 JavaScript 前驱知识，为路由框架学习提供基础知识储备。

第 2 章，讲解 history 库的基础知识及原理，并介绍与 history 库相关的限制及注意事项，帮助读者学习导航基础。

第 3 章，介绍 React v16.3 及之后版本的一些新特性，包括 Context、Hooks 等，旨在帮助读者学习与理解 Router 设计中的相关知识。

第 4 章，首先对 React Router 进行基本的介绍，并对 React Router 的发展历程进行梳理，分析 React Router 的版本演进过程，使读者对 React Router 有一个全方位的了解。之后，从一个简单的工程实例出发，使读者了解 React Router 的简单使用方法，并引出 React Router 三要素，为后面章节做好铺垫。

第 5 章，全面介绍路由系统的第一个基本要素——Router。在对不同种类的 Router 进行介绍的同时，通过源码介绍 Router 的基本原理，并介绍 Router 相关的 Hooks。学完本章，将为理解整个路由系统的原理打好重要的基础。

第 6 章，介绍路由系统的第二个要素——Route。首先介绍 Route 中的两个基本元素：路径与渲染，并介绍 Route 为组件提供的属性与 Route 的相关配置。在通过 Route 源码巩固后，通过相关 Hooks 与多个实战案例，为读者提供丰富的实战经验总结。

第 7 章，介绍路由系统中的导航，包括基本导航组件、带激活态的导航组件等。在介绍基础使用方法的同时，也对源码实现进行分析，帮助读者深入了解导航组件的设计，并通过实战案例强化读者对导航的理解。

第 8 章，介绍帮助组件及方法，恰当地使用相关组件或方法，可以提升开发效率。本章在介绍各组件与方法的基础用法时，也穿插了各组件的源码解析，并通过实战案例，帮助读者深入学习和掌握各组件的设计思路及实际用法，提升读者在实战中的开发效率。

第 9 章，在学好以上各章知识的基础上，读者能清晰地看到路由系统的全貌，且对 React Router 有全面的掌握。以此为基础，本章提供多个进阶案例，更为读者提供组件设计的思路和方法。通过本章系统性的学习，读者可全方位地学习与理解前端路由，并提升对前端路由的整体认识，掌握前端领域路由的设计思路与方法。

本书既能帮助初学者快速上手，又能帮助有一定基础的开发者深入理解 React Router 的设计实现，从而加深对路由系统的理解。读完本书，读者不仅能清楚各场景是怎么使用 React Router 的，更能深入理解 React Router 的设计原理，对 React Router 做出定制化的改造，以面对日渐复杂的页面结构与需求。

阅读本书需要有一定的 React、TypeScript 或 JavaScript 基础。本书适合有 JavaScript 或 TypeScript 基础，想要学习使用 React Router，或对 React Router 实现原理和工程实践感兴趣的开发者阅读。

示例代码：

读者可以从 GitHub 上获取本书的源码：https://github.com/klfzlyt/react-router-tutorial。

本书的完成离不开在各个方面给予我支持和帮助的人，感谢我的同事曾静益、李宏，他们给本书提出了很多宝贵的意见；感谢电子工业出版社博文视点的编辑宋亚东，他在编辑和审校本书期间提出了宝贵的意见；最后，感谢我的家人、朋友、同学在我创作本书期间给予的宽容和支持。

如果您在阅读本书的过程中有任何问题，可以发送邮件到作者的邮箱反馈：klfzlyt@outlook.com。由于作者水平有限，不足之处在所难免，请广大读者批评指正。

李杨韬

目 录
Contents

第 1 章　导航相关 JavaScript 前驱知识 ··············· 1
　1.1　URI 和 URL ··· 1
　　　1.1.1　URI 和 URL 简介 ·································· 1
　　　1.1.2　浏览器 URI 编码 ···································· 3
　1.2　浏览器记录 ··· 4
　　　1.2.1　history.pushState ·································· 5
　　　1.2.2　history.replaceState ································ 9
　　　1.2.3　通过相对路径添加和修改浏览器记录 ·············· 11
　　　1.2.4　在 base 元素存在的情况下添加和修改浏览器记录 ······ 14
　1.3　在浏览器中跳转 ·· 15
　　　1.3.1　window.history.go ································ 15
　　　1.3.2　window.history.forward ·························· 16
　　　1.3.3　window.history.back ····························· 16
　　　1.3.4　window.location.href ···························· 17
　　　1.3.5　window.location.hash ···························· 17
　　　1.3.6　window.location.replace ························· 18
　1.4　浏览器相关事件介绍 ·· 19
　　　1.4.1　popstate 事件 ····································· 19
　　　1.4.2　hashchange 事件 ·································· 21
　　　1.4.3　手动触发事件 ···································· 22
　1.5　小结 ·· 24
　参考文献 ··· 24

第 2 章　history 库详解 ··· 26
　2.1　history 库概述 ··· 26

2.2 browserHistory ········· 32
2.2.1 创建 browserHistory ········· 32
2.2.2 history 导航 ········· 33
2.2.3 history 监听 ········· 37
2.3 hashHisotry ········· 38
2.3.1 创建 hashHisotry ········· 38
2.3.2 history 导航 ········· 40
2.3.3 history 监听 ········· 45
2.3.4 history.createHref ········· 47
2.4 memoryHistory ········· 49
2.4.1 创建 memoryHistory ········· 49
2.4.2 history 导航 ········· 50
2.4.3 history 监听 ········· 53
2.5 history 库原理 ········· 54
2.5.1 history 库的运行流程 ········· 54
2.5.2 history 模拟历史栈 ········· 55
2.5.3 browserHistory 事件处理 ········· 57
2.5.4 hashHistory 事件处理 ········· 58
2.5.5 history.block 原理解析 ········· 60
2.6 history 库限制 ········· 63
2.6.1 history.block 的使用限制 ········· 63
2.6.2 decodeURI 解码问题 ········· 64
2.7 使用 history 替换页面 search 和 hash 示例 ········· 70
2.8 小结 ········· 71
参考文献 ········· 71

第 3 章 React 相关知识 ········· 72
3.1 Context ········· 72
3.2 Hooks ········· 76
3.2.1 useState ········· 76
3.2.2 useEffect ········· 78
3.2.3 useLayoutEffect ········· 84
3.2.4 useRef ········· 86

		3.2.5 useMemo	87
		3.2.6 useContext	89
		3.2.7 自定义 Hook	89
	3.3	Refs	90
		3.3.1 createRef	90
		3.3.2 forwardRef	91
	3.4	Memo	91
	3.5	小结	94
	参考文献	94

第 4 章 认识 React Router ················· 95

4.1	React Router 是什么	95
4.2	React Router 版本的演进	96
4.3	静态路由与动态路由	97
4.4	使用 React Router 实现一个工程应用	98
4.5	小结	107

第 5 章 Router ················· 109

5.1	Router 是什么	109
5.2	Router 源码解析	110
	5.2.1 history 监听	110
	5.2.2 提供初始 Context	110
	5.2.3 提前监听	113
5.3	BrowserRouter	113
5.4	HashRouter	114
5.5	NativeRouter	115
5.6	StaticRouter	116
5.7	相关 Hooks	121
	5.7.1 useRouterContext	122
	5.7.2 useHistory	122
	5.7.3 useLocation	123
5.8	小结	124
参考文献		124

第 6 章 Route ·······125

- 6.1 Route 是什么 ·······125
- 6.2 Route 的两个基本要素 ·······125
  - 6.2.1 Route 的第一个要素：path ·······126
  - 6.2.2 Route 的第二个要素：组件渲染方式 ·······134
- 6.3 Route 传入组件的 3 个参数 ·······138
  - 6.3.1 match ·······138
  - 6.3.2 location ·······140
  - 6.3.3 history ·······141
- 6.4 Route 的其他配置 ·······142
  - 6.4.1 location ·······142
  - 6.4.2 exact ·······142
  - 6.4.3 strict ·······143
  - 6.4.4 sensitive ·······144
- 6.5 Route 源码解析 ·······144
  - 6.5.1 上下文的更新 ·······144
  - 6.5.2 运行流程 ·······147
- 6.6 相关 Hooks ·······149
  - 6.6.1 useRouteMatch ·······149
  - 6.6.2 useParams ·······149
- 6.7 Route 实战案例 ·······150
  - 6.7.1 嵌套 Route ·······150
  - 6.7.2 相对路径 Route ·······152
  - 6.7.3 重定向 Route ·······153
  - 6.7.4 默认子组件 Route ·······156
  - 6.7.5 缓存 Route ·······158
  - 6.7.6 Route 渲染组件的可访问性支持 ·······165
  - 6.7.7 query 及命名参数 ·······166
  - 6.7.8 Route 中的代码拆分 ·······168
- 6.8 小结 ·······169
- 参考文献 ·······170

第 7 章 Link 171
7.1 Link 介绍 171
7.1.1 Link 的定义及属性 171
7.1.2 Link 源码解析 174
7.2 NavLink 176
7.2.1 带激活态的 Link 176
7.2.2 转义特殊字符 178
7.2.3 NavLink 源码解析 178
7.3 DeepLinking 181
7.4 BackButton 182
7.5 导航实战案例 183
7.5.1 为导航组件扩展路由匹配 183
7.5.2 相对上下文路径导航组件 184
7.5.3 相对上下文路径的导航方法 185
7.5.4 为导航组件扩展 search 和 hash 支持 188
7.6 小结 191
参考文献 191

第 8 章 其他路由组件及方法 192
8.1 Switch 192
8.1.1 Switch 简介 192
8.1.2 Switch 源码解析 194
8.2 Redirect 196
8.2.1 基本跳转 196
8.2.2 条件跳转 197
8.2.3 源码解析 198
8.3 Prompt 201
8.4 withRouter 203
8.5 matchPath 205
8.6 实战案例 206
8.6.1 路由动画 206
8.6.2 Prompt 组件 210

		8.6.3　404 页面 ··· 212
		8.6.4　不销毁未命中路径组件的扩展 Switch ····································· 215
	8.7　小结 ·· 218
	参考文献 ·· 219

第 9 章　进阶实战案例 ·· 220

	9.1　路由组件的滚动恢复 ·· 220
		9.1.1　scrollRestoration ··· 220
		9.1.2　容器元素滚动恢复 ··· 221
		9.1.3　滚动管理者 ScrollManager ·· 221
		9.1.4　滚动恢复执行者 ScrollElement ·· 224
		9.1.5　多次尝试机制 ··· 227
	9.2　异步 history 方法 ··· 229
		9.2.1　提升 history 方法 ·· 229
		9.2.2　导航感知 ··· 231
	9.3　为路由引入 hash 定位 ·· 233
		9.3.1　页面加载 ··· 233
		9.3.2　异步数据加载 ·· 235
	9.4　为组件引入路由生命周期 ··· 237
		9.4.1　路由生命周期 ·· 237
		9.4.2　实现路由生命周期高阶组件 ··· 241
	9.5　React Router 状态同步 Redux ·· 246
		9.5.1　接入 connected-react-router ··· 246
		9.5.2　connected-react-router 原理分析 ····································· 249
	9.6　React Router 状态同步 Mobx ·· 251
	9.7　路由与组件的结合实战 ·· 252
		9.7.1　路由结合 Tabs 组件 ··· 252
		9.7.2　路由结合 Modal 组件 ·· 256
		9.7.3　路由结合 BreadCrumb 组件 ·· 257
	9.8　为 history 方法引入前置中间件 ··· 260
		9.8.1　Redux 中间件 ··· 260
		9.8.2　中间件定义 ··· 261
		9.8.3　实现 history 中间件 ·· 263

9.9 组件路由化 268
9.9.1 为组件加入 path 属性 268
9.9.2 为组件赋予路由 269
9.10 路由与页签机制 274
9.10.1 页签介绍 274
9.10.2 页签配置 276
9.10.3 页签实现 277
9.11 在 React Hooks 中使用路由 282
9.11.1 通过 React Hooks 获得路由组件 282
9.11.2 实现 useHookRoutes 284
9.12 微服务路由 286
9.12.1 微服务介绍 286
9.12.2 实现示例 287
9.13 配置化路由扩展 294
9.13.1 配置化路由与 react-router-config 294
9.13.2 重新实现配置化路由 298
9.14 配置化路由综合示例 303
9.14.1 路由配置 304
9.14.2 导航 310
9.14.3 使用页签组件 318
9.14.4 页签栈维护 325
9.15 小结 335
参考文献 336

附录 A 从 React Router v3.x 迁移到 React Router v4.x 及以上版本 337

第 1 章
导航相关 JavaScript 前驱知识

本章介绍相关的基础知识，有一定经验的读者可选择阅读。

1.1 URI 和 URL

1.1.1 URI 和 URL 简介

统一资源标识符（Uniform Resource Identifier，URI），允许用户对网络中的资源通过特定的协议进行交互操作。RFC 2396 文档对 Uniform Resource Identifier 各部分的定义如下。

Uniform：规定统一的语法格式，以方便处理多种不同类型的资源，而无须根据上下文环境来识别资源类型。

Resource：可标识的任何资源。资源不仅可以为单一对象，也可以为多个对象的集合体。

Identifier：表示可标识的对象，也称为标识符。

在一般情况下，URI 为由某个协议方案表示的资源的定位标识符。协议方案是指访问资源时所使用的协议类型名称。HTTP 就是协议方案的一种，除此之外，还有 FTP、file、TELNET 等 30 种标准 URI 协议方案。协议方案由互联网号码分配局（IANA）管理颁布。URI 使用字符串标识某一互联网资源，常用的 URL 作为 URI 的子集，表示某一互联网资源的地点。

URI 的通用语法由 5 个组件组成：

```
URI = scheme:[//authority]path[?query][#fragment]
```

其中，authority 组件可以由以下 3 个组件组成：

```
authority = [userinfo@]host[:port]
```

在 authority 中，userinfo 作为登录信息，通常形式为指定用户名和密码，当从服务器获取资源时作为身份认证凭证使用。userinfo 为可选项。

服务器地址 host 在使用绝对路径 URI 时需指定访问的服务器地址，地址可以为被 DNS 解析的域名，如 example.com，或者 192.168.1.1 的 IPv4 地址及用方括号括起来的 IPv6 地址[0:0:0:0:0:0:0:1]。

port 为服务器连接的网络端口号，作为可选项，如果不指定，则自动使用默认的端口号。

在 URI 语法中，scheme 为协议方案名，在使用 HTTPS 或 HTTP 等协议方案名时不区分大小写，最后一个符号为冒号 ":"。协议方案名也可使用 javascript:、data: 指定脚本程序或数据。

path 为带层次的文件路径，指定服务器上的文件路径，以访问特定的资源。

query 为查询字符串，针对指定路径的文件资源，可使用查询字符串传入任意查询参数。

fragment 为片段标识符，通常标记已获取资源的子资源，为可选项。

RFC 3986 文档列举了几种 URI 示例：

```
ftp://ftp.is.co.za/rfc/rfc1808.txt
http://www.ietf.org/rfc/rfc2396.txt
ldap://[2001:db8::7]/c=GB?objectClass?one
mailto:John.Doe@example.com
news:comp.infosystems.www.servers.unix
tel:+1-816-555-1212
telnet://192.0.2.16:80/
urn:oasis:names:specification:docbook:dtd:xml:4.1.2
https://daisy@example.com:123
```

在以上 URI 示例中，ldap://[2001:db8::7]/c=GB?objectClass?one 的 schema 为 ldap，authority 为 //[2001:db8::7]，path 为 /c=GB，query 为 ?objectClass?one；https://daisy@example.com:123 的 schema 为 https，authority 为 //daisy@example.com:123，其中 userinfo 为 daisy，host 为 example.com，port 为 123。

统一资源定位器（Uniform Resource Locators，URL）作为 URI 的一种，如同网络的门牌，标识了一个互联网资源的"住址"，如 http://www.example.com 表示通过 HTTP 协议从主机名为 www.example.com 的主机上获取首页资源。

URL 的语法定义与 URI 一致：

```
URL = scheme:[//authority]path[?query][#fragment]
```

以 https://example.com:80/foo/baz?title=router 为例，其中 https 表示加密的超文本传输协议，example.com 为服务器的地址，80 为服务器上的端口号，/foo/baz 为资源路径，?title=router 为路径的查询，以 "?" 开头，各个参数以 "&" 分隔，以等号 "=" 分开参数名称与数据。

如无特别说明，本书以 URL 作为网络资源地址的描述。

1.1.2 浏览器 URI 编码

1. 百分号编码

URI 编码使用的是百分号编码（Percent-encoding）。对于需要编码的字符，将其表示为两个十六进制的数字，然后在其前面放置转义字符 "%"，并替换原字符相应位置进行编码。

RFC 3986 文档规定，URI 中只允许包含未保留字符及所有保留字符。其中，未保留字符包含英文字母（a~z，A~Z），数字（0~9），-、_、.、~ 4 个特殊字符，共 66 个。对于未保留字符，不需要进行百分号编码。保留字符是那些具有特殊含义的字符。RFC 3986 文档中规定了 18 个保留字符：

```
!*'();:@&=+$,/?#[]
```

在 URI 中，保留字符有特殊的意义，如 "?" 表示查询，"#" 表示片段标识。如果希望保留字符不表示特定的意义，仅表示一般字符，那么需要对保留字符进行 URL 编码。例如，"/" 作为保留字符具有特殊意义，即作为 URI 的路径分隔符，如果希望其不作为 URI 的路径分隔符，仅作为一般字符，则需编码为 "%2F" 或 "%2f"。

2. encodeURI

encodeURI 是 W3C 的标准（RFC 3986），不对 ASCII 字母和数字进行编码，不对 20 个 ASCII 标点符号（-、_、.、!、~、*、'、(、)、;、/、?、:、@、&、=、+、$、,、#）进行编码。对于 66 个未保留字符，18 个保留字符，除去 2 个不安全的保留字符 "["、"]"，encodeURI 的不编码集为 82 个。

对于非 ASCII 字符，encodeURI 需要将其转换为 UTF-8 编码字节序，然后在每个字节前面放置转义字符（%）进行百分号编码，并置入 URI 中的相应位置。

```
// 不编码
encodeURI("?:@&=+$,#-_.!~*'();/")      // "?:@&=+$,#-_.!~*'();/"
encodeURI("路由")                       // "%E8%B7%AF%E7%94%B1"
// "#" 不编码，则 "#" 对应的 "%23" 也不会解码
decodeURI('/view/%23abc')              // "/view/%23abc"
```

UTF-8：UTF-8 具有无字节序要求、单字节特性节约内存、向后兼容 ASCII、错误兼容性好等优点。一个纯 ASCII 字符串也是一个合法的 UTF-8 字符串，所以现存的 ASCII 文本不需要转换。为传统的扩展 ASCII 字符集设计的软件通常可以不经修改或经过很少修改就能与 UTF-8 一起使用。

3. encodeURIComponent

encodeURIComponent 假定参数是 URI 的一部分（比如协议、主机名、路径或查询字符串），因此，encodeURIComponent 将转义除字母、数字、"("、")"、"."、"!"、"~"、"*"、"'"、"-"和 "_" 外的所有字符。例如，对 "name=va&lu=" 进行 encodeURIComponent 编码后结果为 '"name%3Dva%26lu%3D"'。对于 URL 组成部分中的特殊字符，通常需要使用 encodeURIComponent 进行编码，如：

```
// （、;、/、?、:、@、&、=、+、$、,、#、）都会编码
name=encodeURIComponent('va&lu=')
// "name=va%26lu%3D"
```

使用 encodeURIComponent 的好处是，如果字符串被发送到服务端进行解析，那么服务端会把紧跟在 "%" 后的字节当成普通的字节，而不会把它当成各个参数或键值对的分隔符。

相比 encodeURIComponent，encodeURI 被用作对一个完整的 URI 进行编码，而 encodeURIComponent 则被用作对 URI 的一个组件或者 URI 中的一个片段进行编码。从上面的编码示例来看，encodeURIComponent 编码的字符范围要比 encodeURI 大。

1.2 浏览器记录

浏览器记录是浏览器中各页面用户的导航记录。在现代浏览器中，浏览器记录并没有直接的 API 可获取，其可通过 window.history.length 获取当前记录栈的长度信息。浏览器记录由浏览器统一管理，并不属于某个具体的页面，与页面形式及其内存均无关。

window.history 对象上存在着诸多属性。在浏览器中打开控制台，打印出 history 对象，如下面的 API 列表：

```
interface History {
    readonly length: number;
    scrollRestoration: "auto" | "manual";
    readonly state: any;
    back(): void;
    forward(): void;
    go(delta?: number): void;
    pushState(data: any, title: string, url?: string | null): void;
    replaceState(data: any, title: string, url?: string | null): void;
}
```

从 https://caniuse.com 中的数据来看，约 97%的浏览器支持历史记录管理（Session History Management）的特性，这一特性主要包括 window.history.pushState、window.history.replaceState 与 popstate 事件，它们分别可以添加和修改历史记录条目。一般通常在浏览器的全局对象 window 上操作 history 对象，在明确 history 挂载在全局对象上后，基于此进行约定，下文如无特别说明，history 与 window.history 等价。

1.2.1 history.pushState

1. 基本用法

history.pushState 方法作为 HTML5 特性的一部分，目前被广泛使用。history.pushState 用于无刷新增加历史栈记录，调用 history.pushState 方法可改变浏览器路径。

history.pushState 方法需要 3 个参数：状态对象、标题（目前被忽略）和可选的 URL。其语法格式为

```
pushState(state : Object, title : String, [url : String]) : undefined
```

当设置第三个参数 URL 时，可改变浏览器的 URL，且不会刷新浏览器。

```
window.history.pushState(null,null,'/foo/baz');
console.log(location.pathname);      // "/foo/baz"
```

如果 URL 中包含 Unicode 字符，则浏览器也会将字符按 UTF-8 编码。

```
window.history.pushState(null,null,'/中文');
console.log(location.pathname)       // /%E4%B8%AD%E6%96%87
```

虽然在浏览器地址栏中显示的还是中文字符串，但实际上代码得到的是已经编码过的字符串。

除设置第三个参数为字符串外，在笔者测试的 Chrome 77 版本中同样支持传入 URL 对象。

```
// 如果当前的域名为 https://example.com
window.history.pushState(null,null,new URL('https://example.com/c'));
console.log(location.pathname);        // "/c"
```

history.pushState 方法的第一个参数为需要传入的状态，状态的类型可以为可实施结构化拷贝算法的任意类型。在设置了第一个参数后，可通过 history.state 读取。

```
window.history.pushState(true,null,'/foo/baz');
console.log(history.state); // true
window.history.pushState({a:1},null,'/foo/baz');
console.log(history.state); // {a: 1}
```

因为历史栈由浏览器统一管理，不属于某个具体页面，并不存在于页面的内存中，所以历史栈在刷新页面后不会丢失，栈中记录的各 state 对象也为持久化存储，在导航过程中也不会丢失。

history.pushState 使用结构化拷贝算法进行序列化存储，会将拷贝后的结果记录在历史栈的记录中。结构化拷贝算法除了能拷贝基本类型，还能拷贝更多的对象类型。相比 JSON 的序列化，这样的序列化手段更为安全，如循环引用的对象，结构化序列的手段将会序列化成功，而 JSON 的序列化将会报错，原因在于结构化序列的手段保存了每一个访问过的对象的记录，遇到复制过的对象会进行跳过。对此感兴趣的读者可以参考学习 lodash 的 cloneDeep 方法。结构化拷贝算法要注意特殊场景，如果 history.pushState 的 state 对象中有 dom 节点、error 对象、function 函数等，则调用 history.pushState 方法会抛出异常，且对某些对象的特定属性，如 regExp 的 lastIndex、object 对象的 setter 和 getter 等，结构化拷贝的过程都会丢失。

结构化克隆算法是由 HTML5 规范定义的用于序列化复杂 JavaScript 对象的一个新算法。它比 JSON 更有能力，因为它支持包含循环图的对象的序列化——对象中包含循环引用。此外，在某些情况下，结构化克隆算法可能比 JSON 更高效。

注意，history.pushState 的第一个参数 state，在 Firefox 中有大小限制，超过 640KB 的对象将会抛出异常。history.pushState 的第三个参数 URL 出于安全考虑，需要同源的 URL，例如，当前浏览器的域名为 https://www.github.com，若 history.pushState 的 URL 为 https://stackoverflow.com，则浏览器会抛出 Uncaught DOMException 异常。

2. 历史栈变化

history.pushState 的调用会引起历史栈的变化,浏览器通常会维护一个用户访问过的历史栈,以便用户进行导航。用户通常通过单击浏览器的"前进"和"后退"按钮或者调用 window.history.go 等方法在历史栈中进行移动,可理解为如图 1-1 所示的虚线所表示的栈指针,不改变历史栈的内容,栈内的记录数量不会发生变化。

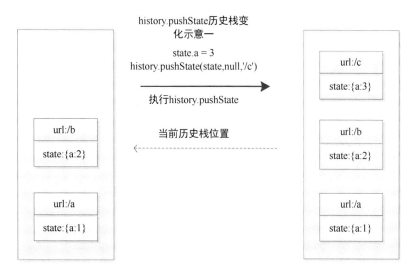

图 1-1　history.pushState 添加历史栈

而当调用 history.pushState 方法时,历史栈的内容会被修改,行为表现为添加历史栈的栈记录,同时也会改变指针指向。如图 1-1 所示,若当前的路径地址为/b,当前的栈指针也指向/b 的位置,在调用 history.pushState({a:3},null,'/c')方法后,则栈记录加 1,栈指针也指向最新的栈记录位置。

```
console.log(history.state);           // {a:2}
// history.length 返回历史栈的记录数量
console.log(history.length);          // 2
console.log(location.pathname);       // /b
history.pushState({a:3},null,'/c');
// 更新了状态
console.log(history.state);           // {a:3}
// 长度加 1
console.log(history.length);          // 3
```

```
// 地址变化
console.log(location.pathname); // /c
```

同时,要注意,如果当前栈指针不在栈顶,如当前栈的数量为 3,单击浏览器的"后退"按钮,使得栈指针指向栈底后,再次调用 history.pushState({a:4},null,'/d')方法,不仅会改变栈指针指向,而且会更新栈的内容,如图 1-2 所示。

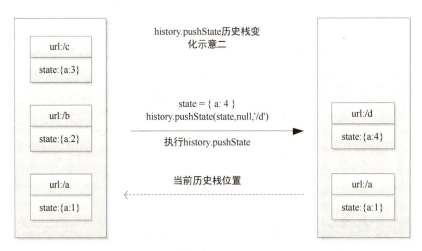

图 1-2　非栈顶情况 history.pushState 更新栈的内容

history.pushState 会在当前指针所指的栈记录后一个位置添加新的历史记录,并使之成为新的栈顶。

```
history.go(-2);
// 当前位于栈底
console.log(history.state); // {a:1}
console.log(history.length); // 3
console.log(location.pathname); // /a
history.pushState({a:4},null,'/d');
// 更新了状态
console.log(history.state); // {a:4}
// 长度变为 2
console.log(history.length); // 2
// 地址变化
console.log(location.pathname); // /d
```

注意,这里减少了栈的记录数量,栈记录数从 3 变为 2,地址/d 成了新的栈顶。

如果不传入第三个参数 URL，则浏览器的地址栏不会发生变化，但是加入一个历史栈，history.length 就会发生相应的变化。

```
history.length //8
history.pushState(null,null)
history.length // 9
```

1.2.2　history.replaceState

1. 基本用法

history.replaceState 的用法与 history.pushState 非常相似，区别在于 history.replaceState 将修改当前的历史记录项而不是新建一个。其语法为：

```
replaceState(state : Object, title : String, [url : String]) : undefined
```

当需要更新当前栈指针所指向的栈记录，而不是增加历史栈时，可使用 history.replaceState 方法，该方法不会使 history.length 发生变化。

```
history.replaceState(null,null,'/foo');
console.log(history.length);  // 数量为 1
history.replaceState(null,null,'/baz');
console.log(history.length);  // 数量为 1
```

同 history.pushState 方法类似，history.replaceState 方法也可传递 state 到历史栈的栈记录中。

```
history.replaceState({a:1},null,'/baz');
console.log(history.state); // {a: 1}
```

对于 state 对象，history.replaceState 同 history.pushState 一样使用结构化拷贝算法，如对象中不能设置函数：

```
history.replaceState({a:function(){}},null,'/baz')
// error: Uncaught DOMException: Failed to execute 'replaceState' on 'History': function(){} could not be cloned.
```

2. 历史栈变化

history.replaceState 不会改变历史栈中记录的数量，如图 1-3 所示，当位于路径/b 时，调用 history.replaceState({a:3},null,'/c')方法会更新当前栈的信息，栈记录/b 会被替换为/c，此时/b 的记录会丢失，栈的记录数量不会发生变化。

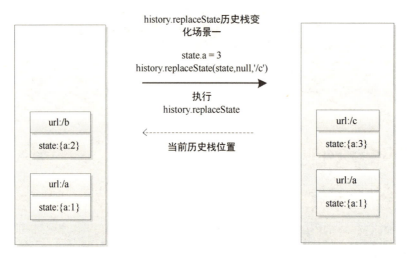

图 1-3　history.replaceState 更新历史栈栈顶记录

```
console.log(location.pathname);      // /b
console.log(history.length);         //历史栈中记录数量为2
history.replaceState({a:3},null,'/c');
console.log(location.pathname);      // /c
console.log(history.length);         // 历史栈中记录数量为2
```

如果当前的栈指针指向栈中间的记录，则此时调用 history.replaceState 方法，仅改变当前栈指针所指向的记录，如图 1-4 所示。

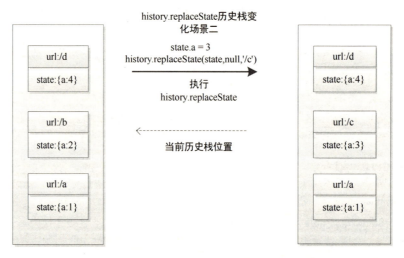

图 1-4　history.replaceState 更新非栈顶记录

/b 的记录被替换为/c 的记录，栈顶的/d 记录不受影响，且栈指针依然位于中间位置。

history.replaceState 与 history.pushState 的主要不同是，history.replaceState 会替换当前指针位置的历史记录，并不会移动指针，也不会入栈新内容，history.length 不会发生变化。

1.2.3 通过相对路径添加和修改浏览器记录

history.pushState 或 history.replaceState，除支持绝对路径导航外，还支持相对路径导航，如：

```
window.history.pushState(null,null,'../one')
window.history.pushState(null,null,'./one/two')
```

熟悉 Node.js 的读者，可参考 url.resolve 相对路径的解决规则。对于相对路径导航，其遵循以下规则。

如果路径以"/"开头，则会替换掉整个路径。

如果路径不以"/"开头，则会得到相对当前 URL 地址的路径（在浏览器无 base 元素存在的情况下），根据路径解决规则会替换 URL 地址中的最后一级目录，即最后一个"/"分隔符后面的路径部分，如：

```
// 当前路径为 /one/two/three
console.log(location.pathname);     // "/one/two/three"
window.history.pushState(null,null,'four')
console.log(location.pathname);     // "/one/two/four"
```

注意，如果当前路径的最后一个字符为"/"，则可认为"/"后紧接空字符串，执行相对路径导航会替换空字符串部分。

```
// 当前路径为 /one/two/
console.log(location.pathname);     // "/one/two/"
window.history.pushState(null,null,"three");
console.log(location.pathname);     // "/one/two/three"
```

如果路径中含有"."".."，则表示当前路径及上一级路径。

```
// 当前路径为 /one/two/three
console.log(location.pathname);     // "/one/two/three"
// 在本级路径更改
window.history.pushState(null,null,'./four')
console.log(location.pathname);     // "/one/two/four"
// 本级路径改为 five，下一级路径为 six
```

```
window.history.pushState(null,null,'./five/../six');
console.log(location.pathname);    // "/one/two/five/six"
```

对于不带具体路径名的操作，可认为其设置最后一级路径为空字符串。

```
// 当前路径为 /one/two/three
console.log(location.pathname);    // "/one/two/three"
// 把当前路径设置为空字符串
window.history.pushState(null,null,'.')
// 或者使用 window.history.pushState(null,null,'./')
console.log(location.pathname);    // "/one/two/"
```

对于不带"."的路径，如 pushState(null,null,'four')，其默认为 pushState(null,null,'./four')。

同时，如果 URL 最后一个字符为"/"，最后一级路径为空字符串，并且当调用 window.history.pushState(null,null,'.')把当前路径也设置为空字符串时，由于最后一级路径都为空字符串，URL 将不发生任何变化。

对于".."操作符，其表明回到上一级路径。

```
// 当前路径为 /one/two/three
console.log(location.pathname);    // "/one/two/three"
// 回到上一级路径 two，再将当前路径 two 设置为 four
window.history.pushState(null,null,'../four');
console.log(location.pathname);    // "/one/four"
// 回到路径 one，并设置路径 one 为/five/six
window.history.pushState(null,null,'../five/six');
console.log(location.pathname);    // "/five/six"
```

同样，如果".."不带具体的路径名，则根据路径解决规则将认为其设置上一级路径为空字符串。

```
// 当前路径为 /one/two/three
console.log(location.pathname);    // "/one/two/three"
// 回到上一级路径 two，并设置路径 two 为空字符串
window.history.pushState(null,null,'..')
// 或者使用 window.history.pushState(null,null,'../')
console.log(location.pathname);    // "/one/"
```

或者

```
// ..回到上一级路径 c，并设置路径 c 为空字符串
window.history.pushState(null,null,'/a/b/c/..')
console.log(location.pathname);    // "/a/b/"
```

当 URL 最后一个字符为"/"，最后一级路径为空字符串时，也遵循同样的规律：

```
// 当前路径为 /one/two/three/
console.log(location.pathname);     // "/one/two/three/"
// 本级路径为空字符串，回到上一级路径 three，并设置路径 three 为 four
window.history.pushState(null,null,'../four');
console.log(location.pathname);     // "/one/two/four"
```

如果路径以"?"或"#"作为跳转 URL 的第一个字符，则浏览器会基于当前的路径地址进行跳转，并会设置对应的参数。

search 情况：

```
// 当前路径为 /one/two/three
console.log(location.pathname);     // "/one/two/three"
history.pushState(null,null,'?name=1')
// 当前 href 路径为 /one/two/three?name=1
console.log(location.pathname);     // "/one/two/four"
console.log(location.search);       // "?name=1"
```

hash 情况，在链接后面拼接：

```
// 当前 href 路径为 /one/two/three?name=1
console.log(location.pathname);     // "/one/two/three"
console.log(location.search);       // "?name=1"
history.pushState(null,null,'#hash')
// 当前 href 路径为 /one/two/three?name=1#hash
console.log(location.hash);         // "#hash"
```

注意，在仅调用 history.pushState 改变 search 的值时，hash 的值会被清理：

```
// 当前 href 路径为 /one/two/three?name=1#hash
console.log(location.pathname);     // "/one/two/three"
console.log(location.search);       // "?name=1"
history.pushState(null,null,'?name=2');
// 当前 href 路径为 /one/two/three?name=2
console.log(location.search);       // "?name=2"
console.log(location.hash);         // "" 空字符
```

如果调用 history.pushState 方法，则需注意此时栈记录的数量依然会改变。

```
history.pushState(null,null,'?name=1')
console.log(history.length)         // 5
```

```
history.pushState(null,null,'?name=1')
console.log(history.length)          // 6
```

1.2.4 在 base 元素存在的情况下添加和修改浏览器记录

如果 HTML 文档中存在 base 元素：

```
<base href="/base/bar">
```

在使用 history.pushState 以 "/" 开头的绝对路径跳转时，base 元素的 href 值是被忽略的，但是如果是相对路径，即路径不以 "/" 开头，则将会使用 base 元素的 href 值作为基准路径，而不使用 window.location.pathname，如：

```
// <base href="/base/bar">
// 当前路径为 /one/two/three
console.log(location.pathname);      // "/one/two/three"
window.history.pushState(null,null,'#hash');
// 将重新以 base 元素的 href 值为基准路径，拼接 hash 字符串
console.log(location.pathname);      // "/base/bar"
console.log(location.hash);          // "#hash"
```

相对路径以 base 元素的 href 值为基准路径：

```
// <base href="/base/bar">
// 当前路径为 /one/two/three
console.log(location.pathname);      // "/one/two/three"
window.history.pushState(null,null,'four');
// 将重新以 base 元素的 href 值为基准路径，替换 base 元素的最后一级路径
console.log(location.pathname);      // "/base/four"
```

对于 query 也同样如此：

```
// <base href="/base/bar">
// 当前路径为 /one/two/three
console.log(location.pathname);      // "/one/two/three"
window.history.pushState(null,null,'?name=1');
// 将重新以 base 元素的 href 值为基准路径，拼接 search 字符串
console.log(location.pathname);      // "/base/bar"
console.log(location.search);        // "?name=1"
```

1.3 在浏览器中跳转

在 Web 浏览器中，在内置的 window 中既存在 history 对象，也包含 location 对象，其中的一些方法可以操作历史记录，如移动栈指针跳转，或者添加栈记录进行导航跳转等。

1.3.1 window.history.go

window.history.go 方法可加载历史列表中的某个特定具体的页面，用来完成在用户历史记录中向后和向前的跳转。其签名为：

```
history.go(number)
```

入参可以是数字，移动历史栈中栈指针的相对距离，"-1"表示后退到上一个页面，"1"表示前进一个页面。在某些浏览器中，也可通过单击"前进"和"后退"按钮旁的历史栈菜单按钮选择需要导航到的页面，这对应于 history.go(1)、history.go(2)、history.go(-2)等。注意，IE 浏览器支持传递 URL 作为参数给 history.go 函数，但这是不标准的。

对于 history.go(0)，其意义为刷新当前页面，与 location.reload 方法行为一致。

```
window.history.go(1)      // 等同于单击浏览器的"前进"按钮
window.history.go(-1)     // 等同于单击浏览器的"后退"按钮
window.history.go(2)      // 前进 2 次，向前移动 2 个页面
window.history.go(0)      // 刷新当前页面
```

对于历史栈指针的移动，如图 1-5 所示。

history.go 方法仅移动栈指针，不会修改栈记录，不会对栈的记录数量造成影响。

history.go 与 history.pushState 相比，history.pushState 并不会造成页面刷新，而 history.go 跳转页面的刷新行为视具体情况而定。history.go 与 history.pushState 的主要区别是，一个不产生历史栈，仅控制栈指针在栈内移动，一个产生历史栈。在行为上，history.go 等同于浏览器的"前进"和"后退"按钮，其调用会触发 popstate 事件；而 history.pushState 会清空当前指针位置之上的所有历史栈，并入栈一个历史记录作为栈顶，同时移动指针指向它。有一个容易产生的误区：在 history.pushState 被调用多次后，单击浏览器的"后退"按钮，浏览器就会完成历史栈的出栈操作。事实上，这时候栈的数量并不会改变。单击浏览器的"前进"和"后退"按钮，包括执行 history.go 等操作都仅是移动指向栈记录的指针，造成指针位置的改变，浏览器并不会进

行出栈。如果需要改变栈的数量，则需要执行 history.pushState。假如一个历史栈有 3 个记录且指针指向栈顶，那么当执行 2 次后退操作后，指针后退到栈底，这时 history.length 还是 3；当再执行一次 history.pushState 时，会产生一个新的栈顶并指向它，history.length 将变为 2，进而才改变了 history.length。注意，history.pushState 永远产生新的栈顶并指向它。

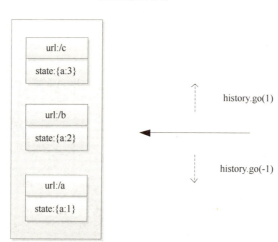

图 1-5　栈指针在栈内移动

window.history.state 作为常量不能直接修改，如 window.history.state={}的赋值将不会成功。

1.3.2　window.history.forward

window.history.forward 作为跳转到当前栈指针所指前一个记录的方法，栈指针向前移动一位。window.history.forward 等同于 history.go(1)，也等同于在浏览器中单击"前进"按钮。注意，前进是否刷新页面取决于历史栈中的栈记录，如果是通过 history.pushState 方法设置的历史栈记录，则调用 window.history.forward 方法不会刷新页面。

1.3.3　window.history.back

在浏览器中向后跳转，window.history.back 使栈指针向后移动一位，这与用户单击浏览器的"后退"按钮效果相同，等同于 history.go(-1)。

1.3.4　window.location.href

使用 window.location.href 导航会产生一个新的历史记录，将字符串设置到 window.location 与设置到 window.location.href 行为一致，如 window.location = 'http://www.example.com'等同于 window.location.href = 'http://www.example.com'。与 history.pushState 产生新记录不同的是，设置绝对路径 URL 到 window.location.href 会刷新页面并重新加载 URL 所指定的内容。

由于 window.location.href 方法增加了历史栈，对原历史栈不造成影响，所以可以单击"后退"按钮返回到之前的历史记录。

```
window.location.href = 'http://a.com'; //
window.location.href = 'http://b.com'; //
window.location.href = 'http://c.com'; //
```

在上面的例子中，单击"后退"按钮将会重新回到 b.com 页面，c.com 的记录不会从历史栈中移除。

window.location.href 同样可使用相对路径进行跳转，需要遵循的规则与 1.2.3 节介绍的一致。

```
console.log(location.pathname); // /a/b/c
window.location.href = 'd';     //
// 页面刷新
console.log(location.pathname); // /a/b/d
```

1.3.5　window.location.hash

window.location.hash 对应 RFC 3986 文档中的 fragment（1.1 节）。hash 以"#"号开头，可通过设置 window.location.hash 改变 URL 的 hash 值。在通过 window.location.hash 设置 hash 值时，"#"号可省略，有时在书写时也可省略 window.。

```
console.log(location.hash);     // ""
location.hash= 'a';             //
// 页面不刷新
console.log(location.hash);     // "#a"
```

改变 hash 同样会产生新的历史栈记录：

```
console.log(location.hash);     // ""
console.log(history.length);    // 1
```

```
location.hash= 'a';                //
console.log(location.hash);        // "#a"
// 栈记录数量发生变化
console.log(history.length);       // 2
```

在设置 location.hash 时要注意，如果设置的 location.hash 值与浏览器 URL 地址的 hash 值相同，就不会触发任何事件，也不会添加任何历史记录。或者如果前后两次对 location.hash 设置了相同的值，则仅第一次 location.hash 设置生效，第二次相同的设置不会产生任何事件和历史记录。

如果希望在改变地址栏 hash 的同时，也不进行入栈操作，则可通过 window.location.replace 实现，如：

```
window.location.replace("#a");
```

类似于 window.history.replaceState 方法，其不会产生新的历史栈记录。

1.3.6　window.location.replace

与 history.replaceState 类似，window.location.replace 会替换当前的栈记录，但在设置绝对路径时，其会刷新页面，重新加载传入的 URL。其语法为：

```
window.location.replace(newURL)
```

与设置 href 不同的是，用 replace 替换了新页面后，旧页面记录不会被保存。在历史栈记录中，这意味着用户将不能用"后退"按钮再次回到旧页面。

```
location.href = 'http://a.com';
location.href = 'http://b.com';
location.replace('http://c.com');
```

在上面的例子中，由于在 b.com 中调用了 location.replace 方法，b.com 的页面记录将消失，替换为 c.com 的页面记录。当用户在 c.com 页面单击浏览器的"后退"按钮时，将回到 a.com 页面。

在某些有表单元素的页面下，用户刷新页面会有 POST 数据的提交，如果想要刷新当前的页面，同时避免 POST 数据的提交，则可以使用：

```
window.location.replace(location.href);
```

对于 window.location.replace，同样可使用相对路径进行跳转，跳转规则与 1.2.3 节中介绍的一致。

1.4 浏览器相关事件介绍

1.4.1 popstate 事件

在 history.pushState 或 history.replaceState 产生的历史栈记录中，当移动栈指针或单击浏览器的"前进"或"后退"按钮时，将触发 popstate 事件，可通过 window.addEventListener 监听该事件。

```
window.addEventListener('popstate', function(event) {
 // do something
});
```

监听函数的参数为对应 popstate 事件的事件对象。对于事件对象 event，event.state 是重点需要关注的，其为移动后对应栈中记录的 state 对象，如图 1-6 所示。

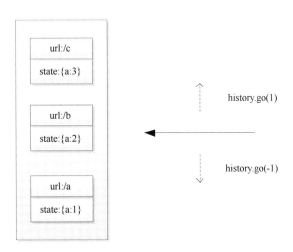

图 1-6　各栈记录保存了 state 对象

当调用 history.go(-1)时，事件回调中得到的 event.state 为{a:1}；当调用 history.go(1)时，事

件回调中得到的 event.state 为{a:3}。

使用 history.pushState 或 history.replaceState 不会触发 popstate 事件。单击浏览器的"后退"或"前进"按钮，或者调用 history 的 go、back、forward 等方法，或者更改部分浏览器的 hash，都会触发此事件。在导航跳转完成后，除了可以从 popstate 事件对象中获取当前的 state 对象，还可以直接从 history 中获取当前的 state 对象，即读取 history.state 变量的值即可。

需要注意的是，由于部分浏览器对 popstate 事件的实现不一致，当网页加载完成后，部分浏览器会触发 popstate 事件，而部分浏览器不会触发 popstate 事件。当编码时，对于这种情况，可使用 history.state 获取状态对象 state，而不是从 popstate 事件对象中获取 state 对象。

注意，在事件中更改 event.state，history.state 不会被改变，如：

```
window.addEventListener('popstate', function(event) {
    event.state.a = 1
});
history.pushState({a:2},null,'/a')
history.pushState({a:3},null,'/b')
history.go(-1)
console.log(history.state) // {a:2} a 不为 1
```

但是直接更改 history.state，将会更改 history.state 的值：

```
console.log(history.state) // {a:1}
history.state.a = 2;
console.log(history.state) // {a:2}
```

在历史栈指针变化后，history.state 将会重新被赋值，即便已经修改过 history.state：

```
console.log(history.state) // {a:1}
history.state.a = 2;
console.log(history.state) // {a:2}
history.pushState({a:3},null,'/b')
console.log(history.state) // {a:3}
history.go(-1)
// 依然为{a:1}
console.log(history.state) // {a:1}
```

由此可知，history.state 同步的是栈记录中的值，每次导航都会获得新的 state 对象。栈记录中的 state 对象是深拷贝存储在浏览器中的，无论在浏览器中进行导航，还是刷新当前页面，抑或是关闭浏览器页签再恢复，历史栈内的内容都存在且不会被销毁。

当前后两次设置相同的 location.hash 值时，不会触发两次 popstate 事件。

```
window.addEventListener('popstate', (event) => {
  console.log("popstate event fired");
});
location.hash = '#/123' // 打印 popstate event fired
console.log(history.length) // 打印 1
location.hash = '#/123' // 没有触发事件
console.log(history.length) // 打印 1
```

若通过 location.href 设置 hash 值，如 location.href ='#/123'，则无论前后设置的值是否相同，都会触发 popstate 事件。当前后两次设置的值相同时，只添加一个历史栈。

```
window.addEventListener('popstate', (event) => {
  console.log("popstate event fired");
});
location.href = '#/123' //打印 popstate event fired
console.log(history.length) // 打印 1
location.href = '#/123' // 打印 popstate event fired
console.log(history.length) // 打印 1
```

1.4.2 hashchange 事件

hashchange 事件用于监听浏览器 hash 值的变化，其监听方式为

```
window.addEventListener('hashchange', function(event) {
    // 触发 hashchange 事件
});
```

hashchange 事件可以通过设置 location.hash、在地址栏中手动修改 hash、调用 window.history.go、在浏览器中单击"前进"或"后退"按钮等方式触发。

在 hashchange 事件的事件响应函数中，可获取事件对象 HashChangeEvent，其除了继承事件对象 Event，拥有 Event 的属性，如 target、type，还提供 oldURL、newURL，分别表示 hash 跳转前的 URL 及 hash 跳转后的 URL。

```
// 假设当前的页面地址为 https://example.com
cosole.log(location.href) // https://example.com
location.hash='foo'
location.hash='baz'
```

```
window.addEventListener('hashchange', function(event) {
    console.log(event.oldURL, event.newURL);
});
// 回退到#foo
history.go(-1)
// 打印 https://example.com#baz "https://example.com#foo"
```

注意，window.history.pushState 不会触发 hashchange 事件，即使前后导航的 URL 仅 hash 部分不同，也是如此，如：

```
window.addEventListener('hashchange', function(event) {
    console.log('hashchange fired')
});
history.pushState(null,null,'/foo#a')
history.pushState(null,null,'/foo#b')
history.pushState(null,null,'/foo#c')
// 以上调用都不会触发 hashchange 事件
```

1.4.3　手动触发事件

对于 popstate 事件，如果调用 history.pushState 方法，则 history.replaceState 方法不会被触发，仅在移动栈指针时才会触发 popstate 事件。对于开发者来说，这缺少一定的控制；不过通过 dispatchEvent 方法，也能实现不移动栈指针便可控制 popstate 事件的触发。调用 dispatchEvent 方法时，dispatchEvent 方法返回值为事件的取消状态，其签名为

```
cancelled = !target.dispatchEvent(event)
```

其接收一个事件实例作为入参，event 是要被派发的事件对象。target 被用来初始化事件和决定将会触发的目标。当该事件是可取消的（cancelable 为 true）且至少一个该事件的事件处理方法调用了 event.preventDefault 时，调用 dispatchEvent 的返回值为 false；否则返回 true。

对于如 window.addEventListener('popstate', handlePopState)监听的 popstate 事件，则可在不移动栈指针的情况下调用 window.dispatchEvent 手动触发。

```
window.dispatchEvent(new PopStateEvent('popstate'))
```

event 事件通过实例化 PopStateEvent 对象得到，并传入事件的 type 为 popstate：

```
window.addEventListener('popstate', ()=>{
    console.log('popstate 触发')
```

```
});
window.dispatchEvent(new PopStateEvent('popstate'))
// 打印 popstate 触发
```

对于 IE 浏览器，PopStateEvent 的类型为 object，不能通过 new 实例化得到对应的 event 实例。event 实例可通过 createEvent 创建，事件类型可通过 initEvent 设置。initEvent 支持 3 个参数：

```
initEvent(eventName, canBubble, preventDefault)
```

这 3 个参数分别表示事件名称、是否可以冒泡、是否阻止事件的默认操作。在 IE 浏览器中，popstate 事件的触发方式为

```
const event = document.createEvent('Event');
event.initEvent('popstate', false, false);
window.dispatchEvent(event);
```

在 IE 浏览器中，Event 类型为对象，而其他浏览器如 Chrome，Event 类型为函数，可兼容写法如下：

```
function dispatchPopStateEvent(){
  if (typeof Event === 'function') {
    window.dispatchEvent(new PopStateEvent('popstate'))
  } else {
    // IE 浏览器的调用
    const event = document.createEvent('Event');
    event.initEvent('popstate', false, false);
    window.dispatchEvent(event);
  }
}
```

同理，对于 hashchange 事件，也可实现对应的手动触发方式：

```
function dispatchHashChangeEvent(){
  if (typeof Event === 'function') {
    // hashchange 事件
    window.dispatchEvent(new HashChangeEvent('hashchange'))
  } else {
    // 在 IE 浏览器下触发
    const event = document.createEvent('Event');
    event.initEvent('hashchange', false, false);
    window.dispatchEvent(event);
```

```
    }
  }
```

对于手动触发的 hashchange 事件，其不受当前 hash 的影响。我们在 1.3 节中介绍过，前后两次设置相同的 location.hash 值不会触发 hashchange 事件，但对于手动触发的方式来说，其可反复触发 hashchange 事件。

```
window.addEventListener('hashchange',(e)=>{
  console.log('hashchanged')
});
dispatchHashChangeEvent();  // 打印 hashchanged
dispatchHashChangeEvent();  // 打印 hashchanged
dispatchHashChangeEvent();  // 打印 hashchanged
```

此方法即使不移动栈指针，也会触发 hashchange 事件，这在某些特定场景中会有具体应用。

1.5 小结

本章主要对导航相关 JavaScript 前驱知识进行了介绍，包括 URI 和 URL、浏览器记录、在浏览器中跳转、浏览器相关事件介绍等。其中，产生或修改浏览器记录的 history.pushState、history.replaceState 原生方法将作为 React Router 中导航的基础方法，并被 React Router 频繁使用；浏览器相关事件，如 popstate 事件，同样是 React Router 中的基础部分。学习本章基础知识，将为学好 React Router 打好重要基础。

参考文献

[1] 上野宣. 图解 HTTP[M]. 于均良，译. 北京：人民邮电出版社，2014.

[2] https://tools.ietf.org/html/rfc3986.

[3] https://developer.mozilla.org/zh-CN/docs/Web/JavaScript/Reference/Global_Objects/decodeURI.

[4] https://zh.wikipedia.org/wiki/%E7%BB%9F%E4%B8%80%E8%B5%84%E6%BA%90%E6%A0%87%87%E5%BF%97%E7%AC%A6.

[5] https://zh.wikipedia.org/wiki/%E7%BB%9F%E4%B8%80%E8%B5%84%E6%BA%90%E5%AE

%9A%E4%BD%8D%E7%AC%A6.

[6] https://zh.wikipedia.org/wiki/%E7%99%BE%E5%88%86%E5%8F%B7%E7%BC%96%E7%A0%81.

[7] https://www.cnblogs.com/jerrysion/p/5522673.html.

[8] https://developer.mozilla.org/zh-CN/docs/Web/API/Window/popstate_event.

[9] https://developer.mozilla.org/zh-CN/docs/Web/API/History_API.

[10] https://developer.mozilla.org/zh-CN/docs/Web/API/Window/location.

[11] https://developer.mozilla.org/zh-CN/docs/Web/API/EventTarget/dispatchEvent.

第 2 章
history 库详解

在 React Router 的依赖列表中，有着一个重要的依赖库——history。如果在项目依赖中没有 history 库，则在安装 React Router 的同时也会安装其依赖列表中的 history 库。React Router v5.1.2 依赖的 history 库版本号为 4.9.0。对于 history 库，也可以单独进行安装：

```
npm install history
```

history 库在 React Router 中扮演着导航执行者与监听者的重要角色。对于 React Router，所有的"副作用"都由 history 库完成。history 库负责与外界的沟通，仅关心路由逻辑的 React Router 不关注外界的情况，需要 history 库为其提供支持。

history 库作为与外界沟通的桥梁，提供了 3 类历史对象，分别是 browserHistory（浏览器历史对象）、hashHistory（哈希历史对象）和 memoryHistory（内存历史对象）。各历史对象负责与不同的外界环境进行沟通，并统一为 React Router 提供支持，它们有着相对一致的接口和复用的代码实现。

本章将介绍各历史对象的对外接口、常用方法与内部源码实现，并分析源码实现原理，同时介绍目前 history 库存在的问题与一些对应的解决办法，旨在帮助读者理解与使用 history 库。

2.1　history 库概述

history 库为 React Router 提供了 history 对象，history 对象提供了相关的地址监听、跳转方法等。history 对象在 React Router 中负责与"外部"的 URL 交互，其在 React Router 中发挥着非常重要的作用。

history 库提供了 3 类历史对象的创建工厂函数，分别是 createBrowserHistory、createHashHistory 和 createMemoryHistory。

history 对象在 React Router 中通常为单例形式，创建 history 对象的方法非常简单，通常为调用相关的工厂 create 方法。

```
import { createBrowserHistory } from "history";
const history = createBrowserHistory();
```

此外，createHashHistory 方法和 createMemoryHistory 方法也可以创建 history 对象。由 createBrowserHistory 方法与 createHashHistory 方法创建的 history 对象依赖于浏览器运行环境；而调用 createMemoryHistory 方法创建的 history 对象可独立于浏览器，在内存中运行。

3 类历史对象都提供了一致的能力与接口，除了通用能力，各历史对象对不同的外界环境也有着不同的能力。首先是通用能力部分，各历史对象都具备：

- 监听外界地址变化的能力。
- 获取当前地址的能力。
- 增加和修改历史栈的能力。
- 在栈中移动当前历史栈指针的能力。
- 阻止跳转，以及定义跳转提示的能力。
- 获取当前历史栈长度及最后一次导航行为的能力。
- 转换地址对象的能力。

其次，各历史对象也有着自身独有的能力。

（1）browserHistory：

- 可提供基准地址。
- 可设置强刷模式。
- 历史栈标志符长度（keyLength）控制。

（2）hashHistory：

- 可设置不同的 hash 模式（hashType）。
- 可提供基准地址。

（3）memoryHistory：

- 访问整个历史栈的能力。
- 设置初始历史地址及初始栈指针的能力。
- 获取当前栈指针的能力。

- 历史栈标志符长度（keyLength）控制。
- 导航确认能力（canGo）。

值得一提的是，历史栈标志符长度控制与提供基准地址的能力由于使用较少，可能在未来某个 history 版本中被移除。

对于 browserHistory、hashHistory 及 memoryHistory，它们的通用能力可用同样的 history 对象类型进行描述。

```typescript
export interface History<HistoryLocationState = LocationState> {
    // 历史栈长度
    length: number;
    // 最后一次导航的导航行为
    action: Action;
    // 当前历史地址
    location: Location<HistoryLocationState>;
    // 添加历史栈记录
    push(path: Path, state?: HistoryLocationState): void;
    // 添加历史栈记录重载方法
    push(location: LocationDescriptorObject<HistoryLocationState>): void;
    // 修改历史栈记录
    replace(path: Path, state?: HistoryLocationState): void;
    // 修改历史栈记录重载方法
    replace(location: LocationDescriptorObject<HistoryLocationState>): void;
    // 移动栈指针
    go(n: number): void;
    // go(-1)
    goBack(): void;
    // go(1)
    goForward(): void;
    // 阻止导航行为
    block(prompt?: boolean | string | TransitionPromptHook): UnregisterCallback;
    // 监听地址变化
    listen(listener: LocationListener): UnregisterCallback;
    // 地址对象转换
    createHref(location: LocationDescriptorObject<HistoryLocationState>): Href;
}
```

对于创建历史对象的工厂方法，createBrowserHistory、createHashHistory 及 createMemoryHistory 都可以传入配置对象，配置对应的历史对象的行为。历史对象都可以传入 getUserConfirmation 跳转确认函数：

```
import { createBrowserHistory } from "history";
const history = createBrowserHistory({
  getUserConfirmation:getUserConfirmation(message, callback){/**/}
});
```

getUserConfirmation 跳转确认函数通常用于在用户操作流程没有结束又产生导航时提醒用户。对于 getUserConfirmation 跳转确认函数，其需要与 history.block 配合使用。history.block 不与具体历史对象相关，对 browserHistory、hashHistory 和 memoryHistory 都通用。

history.block 接受不传入参数，或者传入一个 string、boolean 类型的参数或一个 prompt 函数。当调用 history.block 时，任何导航包括浏览器的前进或后退行为都将被阻止，或者以某种形式提示用户确认导航。此时，内容区域不会发生变化。如果需要解绑 history.block，则可以调用 history.block 返回的注销函数。当调用注销函数后，history 将不会阻止跳转行为。history.block 的函数签名如下，返回一个注销函数：

```
block(prompt?: boolean | string | TransitionPromptHook): UnregisterCallback;
```

当 prompt 为 boolean 类型时，若为 true，则不会阻止路由跳转；若为 false，则阻止路由跳转。当 prompt 为 string 类型时，默认弹出系统的 prompt，一般为 window.confirm。如果创建 history 时传入了 getUserConfirmation，则使用传入的 prompt。如果弹出系统的 prompt，则会根据弹框返回结果进行路由控制，当在弹框中选择"确定"按钮时，则不进行阻止，按路由跳转预期执行；当在弹框中选择"取消"按钮时，则会阻止路由跳转。如下面示例所示：

```
function Example(props){
  const { history } = props;
  // 使用 useRef 保存注销函数
  const historyUnBlockCb = React.useRef<UnregisterCallback>(()=>{});
  React.useEffect(()=>{
    return ()=>{
      // 在组件销毁时调用，注销 history.block
      historyUnBlockCb.current()
    }
  },[]);
  function block() {
    // 解除之前的阻止
```

```
            historyUnBlockCb.current();
            // 重新设置弹框确认，更新注销函数，单击"确定"按钮则正常跳转；单击"取消"
            // 按钮则跳转不生效
            historyUnBlockCb.current = history.block('是否继续？')
        }
        return <>
            <button onClick={block}>阻止跳转</button>
            <button onClick={()=>{
                historyUnBlockCb.current();
            }}>解除阻止</button>
        </>
    }
```

在上例中，若单击"阻止跳转"按钮，则在调用 history.block 函数后，所有的跳转导航都会提示"是否继续？"，如图 2-1 所示。只有当调用 history.block 返回的解除函数后，才会解除跳转提示。注意，上例使用了 React Hooks，关于 React Hooks 部分将在第 3 章介绍。

图 2-1　跳转确认弹框提示

在 history.block 参数列表中，允许传入 prompt 函数，prompt 函数的类型为 TransitionPromptHook：

```
(location: Location, action: Action) => string | false | void;
```

参数列表中的第一个参数为要导航到的 location 地址；第二个参数为调用导航时对应的 action，为 "PUSH"、"REPLACE" 或 "POP"。

当传入的 prompt 函数为 TransitionPromptHook 类型时，可以对跳转阻止与否进行判断，最终将根据函数的返回值情况决定。当函数返回 false 时，将阻止跳转；当函数返回 undefined 或 true 时，将不进行阻止。如果返回的是字符串，则默认使用 window.confirm 进行弹框确认，弹框的提示内容即返回的字符串；同样，如果创建 history 时传入了 getUserConfirmation，则将使用传入的 getUserConfirmation 进行提示。当使用默认的 window.confirm 进行弹窗提示时，若单击"确定"按钮，则进行正常导航；若单击"取消"按钮，则阻止跳转。对于 getUserConfirmation，其类型为：

```
    getUserConfirmation(message: string, callback: (result: boolean) => void): void;
```

当调用 callback(true) 时，将允许导航跳转；当调用 callback(false) 时，导航跳转将不会生效。

在 browserHisotry 和 hashHistory 默认提供的 getUserConfirmation 跳转确认函数内部，都使用了 window.confirm 进行弹窗确认，其实现的 getUserConfirmation 为：

```
export function getUserConfirmation(message, callback) {
  callback(window.confirm(message));
}
```

也可自行定义弹框，如：

```
// message 为 Prompt 组件上的 message 或 message 函数返回的字符串
function getUserConfirmation(message, callback) {
  Modal.confirm({
    title: "跳转确认",
    content: message,
    // 允许用户跳转
    onOk: () => callback(true),
    // 不允许跳转
    onCancel: () => callback(false)
  });
}
```

对于 3 类历史对象通用的 history.action，其意义为最后一次导航的导航行为，其值主要有 "PUSH"、"REPLACE" 和 "POP"。"PUSH" 表示上次导航行为是入栈行为，即增加历史栈记录。"REPLACE" 表示上次导航行为是替换历史栈行为，即修改历史栈记录。"POP" 为调用函数 history.go、history.goBack、history.goForward 或者单击浏览器的"前进"或"后退"按钮时对应的导航行为，类似于 popstate 事件。其特点是仅移动栈指针，不对栈的记录产生影响，其导航行为也应为 "POP"。关于栈指针的移动可查看 1.3 节。

history.length 表示历史栈的长度，通常与 window.history.length 一致。

browserHisotry 与 hashHistory 都有的 basename 为 history 实例的基准路径。当创建 history 传入 basename 时，最终生效的路径为 basename 加上函数中的导航路径。要注意，从 history 得到的路径中不包含 basename，basename 仅生效于浏览器真实 URL 中。以 browserHistory 为例：

```
// 创建基准路径 history
const history = createBrowserHistory({ basename: "/base/foo" });
history.push('/baz');
// 地址栏路径
console.log(window.location.pathname);        // "/base/foo/baz"
```

```javascript
// history 对象中的 pathname
console.log(history.location.pathname);    // "/baz"
```

对于浏览器路径 location，也会剥除 basename 部分：

```javascript
window.history.replaceState(null, null, '/prefix/pathname');
const history = createBrowserHistory({ basename: '/prefix' });
console.log(history.location.pathname);    // '/pathname'
```

注意，basename 会忽略大小写：

```javascript
// 在浏览器中为大写
window.history.replaceState(null, null, '/PREFIX/pathname');
// 创建的 history 为小写
const history = createBrowserHistory({ basename: '/prefix' });
console.log(history.location.pathname);    // '/pathname'
```

2.2　browserHistory

2.2.1　创建 browserHistory

browserHistory 也叫浏览器历史对象，特点为其 location 的 pathname、search 等与浏览器中的 window.location 对象的各属性完全兼容。在现代浏览器中，使用 browserHistory 可获得最大的开发兼容性。由于开发兼容性良好，在 Web 浏览器场景中，browserHistory 应作为首要考虑的 history。创建 browserHistory 的方式为：

```javascript
import { createBrowserHistory } from 'history'
const history = createBrowserHistory();
// 与 window.location 各属性兼容
history.location.pathname === window.location.pathname // true
history.location.search === window.location.search // true
```

对于 createBrowserHistory，也可传入如下 history 配置：

```typescript
export interface BrowserHistoryBuildOptions {
  basename?: string;
  forceRefresh?: boolean;
  // 通用配置在 2.1 节中有过阐述
  getUserConfirmation?: typeof getUserConfirmation;
```

```
  keyLength?: number;
}
```

对于 browserHistory，默认的跳转不会造成页面刷新，如果设置 forceRefresh 为 true，则在跳转过程中会强制刷新页面。

keyLength 表示历史栈中栈记录的 key 字符串的长度，默认为 6。

如果在创建 history 的时候传入了 basename，则通过 createHref、history.push 和 history.replace 等方法都会得到 basename 与 path 的拼接。

```
const history = createBrowserHistory({ basename: '/base' });
const href = history.createHref({
  pathname: '/path',
  search: '?the=query',
  hash: '#the-hash'
});
```

上例中的 href 将为 /base/path?the=query#the-hash。

2.2.2 history 导航

1. history.push

history.push 作为 2.1 节介绍过的历史对象的通用方法，类似于 1.2 节中的 history.pushState，其主要作用为添加一个历史记录。其签名如下：

```
export interface History<any> {
push(path: string, state?: any): void;
push(location: {
    pathname?: string;
    search?: string;
    state?: any;
    hash?: string;
}): void;
}
```

当第一个参数为字符串时，即真实导航路径，底层会使用 history.pushState 方法，无刷新改变 URL；第二个参数对应为 history.pushState 的 state 对象，能持久化地存储状态在浏览器中。注意，state 需要为可结构化克隆的对象，在 1.2 节中有过介绍。

```
history.push('/foo#name')
console.log(history.location.pathname) // /foo
console.log(window.location.pathname) // /foo
history.push('/baz',{a:1})
console.log(history.location.pathname,history.location.state) // /baz, {a:1}
console.log(window.location.pathname,window.history.state.state) // /baz, {a:1}
```

若第一个参数为路径描述对象，则可将 state、hash 等值传入其中。

```
browserHistory.push({
  pathname:'/baz',
  search:'?name=2',
  hash: "#c",
  state: { a:1 }
});
// 等价于
// push(`${location.pathname}${location.search}${location.hash}`, location.state);
```

传入路径描述对象也等价于各部分字符串的拼接。

注意，对于 browserHistory 的每次 push 调用，都将产生一个随机的 key 值，该 key 值可从 browserHistory.location 中获取，并持久化存储于 window.history.state 中。这个随机值的字符串长度由创建 history 时的 keyLength 配置进行控制，默认为 6，其作用为标识本次导航，将在 2.5.2 节介绍。

```
browserHistory.push('/baz');
console.log(browserHistory.location.key); // wewq2e 随机值
console.log(window.history.state.key); // wewq2e 随机值
```

对于 push 方法，在 history 源码中的调用为 pushState 方法。

```
// globalHistory 为原生 window.history, title 部分为 null, href 为跳转到的 URL
// state 部分为 key 属性与当前 push 参数的 state 部分，内部 key 值将标识一次唯一的导航
globalHistory.pushState({ key, state }, null, href);
```

与原始的 pushState 方法不同的是，history.push 方法的行为可在 history.block 被调用后改变，如：

```
history.block();
// 在调用 history.block 后，此 push 调用不产生任何效果
history.push('/foo');
// 跳转弹框确认
history.block('确定是否跳转到/foo?');
```

```
// 调用 push 默认弹出系统弹框，单击"确定"按钮后才进行跳转
history.push('/foo');
```

与 history.pushState 不同的是，history.pushState 函数不会触发 popstate 事件，history.pushState 函数没有对应的回调监听；而没有被阻止的 history.push 方法会在状态更新后，触发 history.listen 监听的回调函数。回调函数的参数为当前最新的地址对象和值为"PUSH"的导航行为标识。

对于 push，其都支持相对路径与仅带参数的调用方式，具体解析规则与 1.2.3 节中介绍的解析规则一致。

```
history.push('../foo');        // 回到上一级路径，并设置为/foo
history.push('./foo');         // 本级路径设置为/foo
history.push('../../bar');     //回到上二级路径，并设置为/bar
history.push('?bar');          // 本级路径不变，设置query
history.push('#foo');          // 本级路径不变，设置hash
```

注意，当创建 history 的 forceRefresh 为 true 时，框架将不使用 pushState 原生方法，而是直接调用 window.location.href = href 刷新页面。

2. history.replace

与 history.push 一样，history.replace 也可以改变浏览器地址。其声明为：

```
replace(path: Path, state?: HistoryLocationState): void;
replace(location: LocationDescriptorObject<HistoryLocationState>): void;
```

在调用 history.replace 时，产生的 action 为"REPLACE"。在使用上，history.replace 的入参与 history.push 的入参类型均一致，既允许传入 state 对象，也可传入相对路径，同样可被 history.block 阻止。

```
history.replace('../foo',{a:1});      //替换当前栈记录为上一级路径/foo
console.log(history.length);          // 1
const unblock = history.block();      // 阻止跳转
history.replace('/baz');              //不会生效
console.log(history.length);          // 1
unblock();                            // 取消阻止
history.replace({
  pathname: '/baz',
    search: '?name=1',
}); // 替换栈记录为 /bar?name=1
console.log(history.length); // 1
```

与 history.push 不同的是，history.replace 用于替换当前栈指针所指记录。在 history 源码内部，history.replace 使用了 history.replaceState：

```
// 内部产生的 key 值将标识一次唯一的导航
globalHistory.replaceState({ key, state }, null, href);
```

在 1.2.2 节中曾介绍过 history.replaceState，而 history.push 方法使用的是 window.history.pushState，其会添加栈记录。对于 history.replace，由于 history.replaceState 的调用不会触发 popstate 事件，所以 history.replace 的调用也不会触发 popstate 事件。

如果初始化的 history 的 forceRefresh 为 true，则 history.replace 将使用 window.location.replace(href)进行页面更新，强制刷新页面。

history.replace 方法的 key 生成与 history.push 方法一致，调用 history.replace 后在 location 中将有一个随机 key 值，key 值同样会持久化存储到 window.history.state 中。

```
history.push('/foo');
console.log(history.location.key)        // qrvfez 随机值
history.replace('/baz')
console.log(history.location.key)        //zvwqwe 随机值
console.log(window.history.state.key)    //zvwqwe 随机值
```

key 值将标识一次唯一的导航，可作为导航的唯一凭证。browserHistory 在导航过程中产生的所有 key 值将被保存在内存中，用于标识栈记录，相关内容将在 2.5.2 节进行介绍。

3. history.go

history.go 及 history.goBack、history.goForward 使用了 window 中的 window.history.go 方法，其仅做了简单的包装：

```
function go(n) {
  globalHistory.go(n);
}
function goBack() {
  go(-1);
}
function goForward() {
  go(1);
}
```

在 1.3.1 节中介绍过 window.history.go 等方法，其仅移动浏览器的栈记录指针，不对栈的内容产生影响，等同于浏览器的前进或后退操作。

当调用 history.go 等方法时，浏览器中监听的 popstate 事件回调函数会触发。browserHistory 会监听此事件更新 location 地址，相关内容将在 2.5.3 节进行介绍。

2.2.3　history 监听

browserHistory 的 browserHistory.listen 方法监听 browserHistory 的 location 的改变，通过 hisotry.push、history.replace、history.go 等方法改变 location，都会触发回调监听函数。各方法除了能获得最新导航的 location 地址，还能获取到不同的跳转类型 action。其调用方式如下：

```
const history = createBrowserHistory();
history.listen((location,action) => {
  // action 为跳转类型
  // 监听 broswerHistory 地址的改变
  console.log(location.pathname,action)
});
history.push('/one')         //打印 /one PUSH
history.push('/two')         //打印 /two PUSH
history.replace('/three')    //打印 /three REPLACE
// 下面的调用等同于调用 history.goBack 或者单击浏览器的 "后退" 按钮
history.go(-1) //打印 /one POP
// history.push 方法的 action 为 "PUSH"，history.replace 方法的 action
// 为 "REPLACE"，其余导航产生的 action 都为 "POP"
```

history.listen 的回调函数仅在产生导航、location 发生变化时被触发，页面初始加载不会被触发。在页面初始加载后，页面地址会同步到 history.location 中。注意，这里监听 location "变化" 并不表示前后导航过程中 location 的值不一致发生变化，当前后导航过程中产生的 location 一致时，同样会触发 history.listen 的回调函数。

如果希望在 history.listen 的回调函数中获取上一次浏览记录，则可通过声明外部变量实现：

```
// 外部变量保存地址
let previousLocation;
history.listen(nextLocation => {
  console.log(previousLocation);
  /* ...
    如判断前后 location 相等与否
    if(locationsAreEqual(previousLocation,nextLocation)){}
```

```
                在相关逻辑处理完后，再更新 previousLocation
             */
    previousLocation = nextLocation;
});
```

2.3 hashHisotry

2.3.1 创建 hashHisotry

　　hashHistory 在浏览器中使用，在不兼容 pushState 等原生接口的浏览器中，可使用 hashHistory 管理地址导航。hashHistory 的特点是路径的所有部分都在浏览器地址的 hash 中，即"#"号之后，读取 window.location.hash 得到的将是 hashHistory 管理的整个地址信息。

　　如果不希望在页面切换时刷新页面，同时希望将页面的 URL 存储在浏览器地址的 hash 中，则可使用 hashHistory。创建 hashHistory 的方式如下：

```
import { createHashHistory } from 'history';
const history = createHashHistory();
```

　　在创建 hashHistory 时，除了能传入 getUserConfirmation、basename，还可设置 hashHistory 的 hash 类型，其配置如下：

```
// 3 种支持类型
export type HashType = 'hashbang' | 'noslash' | 'slash';
export interface HashHistoryBuildOptions {
  basename?: string;
  // hash 类型
  hashType?: HashType;
  // 通用配置在 2.1 节中有过阐述
  getUserConfirmation?: typeof getUserConfirmation;
}
```

　　注意，hashHistory 不接受 keyLength、forceRefresh 的创建配置，这意味着其缺少了 hashHistory.location 中的 key 与强刷模式。

　　getUserConfirmation 与 basename 在 2.1 节中有过介绍，分别是配置弹窗确认函数及配置基准地址。

　　从 TypeScript 类型 HashType 中可以看到，hashType 支持 hashbang、noslash 及 slash 3 种类

型，如果在创建时不指明 hashType，则默认的 hashType 为 slash。

```
const history = createHashHistory();
// 等价于
const history = createHashHistory({
  hashType: 'slash' // 默认类型
});
```

对于 slash 类型的 hashHistory，其 hash 符号 "#" 后都将跟上 "/"，如：

```
history.push('/foo/baz');
console.log(window.location.hash);
// 将得到
// #/foo/baz
```

若创建的 hashType 为 noslash 类型：

```
const history = createHashHistory({
  hashType: 'noslash'
});
```

则 "#" 后没有 "/"：

```
history.push('/foo/baz');
console.log(window.location.hash);
// 将得到
// #foo/baz
```

若创建的 hashType 为 hashbang 类型：

```
const history = createHashHistory({
  hashType: 'hashbang'
});
```

则对应的 hash 路径 "#" 后会接上 "!" 与 "/"：

```
history.push('/foo/baz');
console.log(window.location.hash);
// 将得到
// #!/foo/baz
```

hashbang 是 UNIX 操作系统中的一种写法，名称叫作 Shebang 或 hashbang。hashbang 是一个由 "#" 和 "!" 构成的字符序列 "#!"，其出现在文本文件的第一行的前两个字符。在 Web 应用中，"#" 后面的内容默认不会被爬虫爬取到。近年来，页面 hash 有着与业务相关的含义，不同 hash 对应的网页内容也有所不同。例如路由地址，为了有效地区别这种情况，让搜索引擎更好地抓取数据，Google 提出的解决方案是用 "#!" 进行区分，即 hashbang。当网页爬虫遇到类似的带有 hashbang 的 URL 时，就会对 URL 中的 hash 进行抓取，从而获得更全面的信息。

注意，即使在创建 hashHistory 时没有进行任何操作，路径也可能发生变化。例如在浏览器中输入 https://example.com/，在 createHashHistory 被调用后，默认的 hashType 为 slash 时，浏览器 URL 会变为 https://example.com/#/。这是因为 createHashHistory 在被调用时，会进行一次准备工作：

```
const path = getHashPath();
const encodedPath = encodePath(path);
if (path !== encodedPath) replaceHashPath(encodedPath);
```

由于 encodedPath 变量为 "/"，与 path 变量为空字符串不相同，因而会执行一次 replaceHashPath('/')。执行该操作的目的是初始化 hashType，如果 hashType 为 hashbang，则路径为 https://example.com/#!/。

2.3.2　history 导航

1. history.push

history.push 类似于 1.2.1 节中的 history.pushState，会添加新的历史栈记录。其调用语法为：

```
push(path: string, state?: HistoryLocationState): void;
push(location: LocationDescriptorObject<HistoryLocationState>): void;
```

其接收字符形式的路径，也接收 location 地址描述对象。

hashHistory 使用浏览器地址的 hash 部分作为路由存储，每次 hashHistory.push 调用仅改变浏览器地址的 hash 值，这在某些不支持 HTML5 特性的旧式浏览器中可被使用。hashHistory.push 的底层 history 源码为：

```
function pushHashPath(path) {
  window.location.hash = path;
}
```

调用 window.location.hash 进行一次 hash 改变，从而实现改变浏览器地址 hash 并入栈的操作。

```
const hashHisotry = createHashHistory();
console.log(location.href); // "https://example.com/a/b/c#/"
hashHisotry.push('/baz');
console.log(location.href); // https://example.com/a/b/c#/baz
```

```
hashHisotry.push('/faz?name=1');
console.log(location.href); // https://example.com/a/b/c#/faz?name=1
// location.pathname 部分都不变化
```

注意，对于 location，history.push 调用是按照整个浏览器协议进行的，如可调用 history.push('/foo/baz#121')。这会在浏览器中产生两个#号，第一个#号后为 hashHistory 所标识的地址，第二个#号后为 hashHistory 中 location 的 hash 值，如：

```
const hashHisotry = createHashHistory();
console.log(location.href); // "https://example.com/a/b/c#/"
hashHisotry.push('/baz#foo');
console.log(window.location.href); // https://example.com /a/b/c#/baz#foo
console.log(hashHisotry.location.hash); // #foo
console.log(window.location.hash); // #/baz#foo
```

如同 browserHistory 一样，hashHistory 也可调用 hashHistory.block 阻止导航；导航时也可以传入相对路径参数。

```
hashHisotry.block();
 // 此调用不产生任何效果
hashHisotry.push('/foo');
 // 跳转弹框确认
hashHisotry.block('确定是否跳转到/baz?');
 // 此调用默认使用系统弹框，在单击"确定"按钮后才进行跳转
hashHisotry.push('/baz');
 // 相对本级路径跳转到 biz
hashHisotry.push('./biz');
```

注意，如果 hashHistory.push 设置了 state 参数，则第 4 版本的 history 库会给出警告：

```
Warning: Hash history cannot push state; it is ignored.
```

由于考虑兼容性问题，第 4 版本的 history 库没有使用 pushState 的底层接口。因此，对应的历史栈状态也无法存储，history 进行了忽略并提示开发者。

对于 hashHistory.push 的参数，也可以为 location 对象，其类型描述如下：

```
export interface LocationDescriptorObject<S = LocationState> {
    pathname?: Pathname;
    search?: Search;
    state?: S;
    hash?: Hash;
```

```
    key?: LocationKey;
}
```

若 hashHistory.push 方法的第一个参数为对象，则可分别设置 hashHistory 的 pathname、search 和 hash 等：

```
const hashHisotry = createHashHistory();
console.log(location.href); // "https://example.com/a/b/c#/"
hashHisotry.push({
  pathname:'/baz',
  search:'?name=2',
  hash: "#c"
});
console.log(location.href); // https://example.com/a/b/c#/baz?name=2#c
```

如果希望 hashHistory 也能传递一个 state 对象，则可在 location 对象中设置 state：

```
import { createHashHistory } from 'history';
const hashHisotry = createHashHistory();
hashHisotry.push({
  pathname: '/home',
  search: '?the=query',
  // 注意，不生效于 window.history.state
  state: { some: 'state' }
});
```

第 4 版本的 history 库不建议在 hashHistory 中使用 state，虽然可通过 location 对象传递 state，但是其作为页面级别的 state，不具备持久化 state 的能力。这时，仅能从 hashHistory.location.state 中读取到对象{ some: 'state' }，而不能从 window.history.state 中读取到。例如在浏览器中执行一次后退再前进的操作，由于 window.history.state 没有存储状态，这时读取 hashHistory.location.state，读取到的值将为空。而 browserHistory 可以再次读取到 state 值，所以需要注意，hashHistory.location.state 在导航过程中并不能如 browserHistory 一样其 state 值能得到再现。由于 pushState 等 HTML5 接口已经被广泛使用，在 history 库未来的第 5 版本中将使用 pushState 来模拟 hashHistory，因此持久化的 state 设置会得到支持。此时，hashHistory 设置的 state 也可从 window.history.state 中读取到。相关的持久化能力可查看 1.2.1 节。

2. history.replace

使用 history.replace 可替换历史栈中的栈记录。history.replace 的使用方式与 history.push 类似，其签名如下：

```
replace(path: Path, state?: HistoryLocationState): void;
replace(location: LocationDescriptorObject<HistoryLocationState>): void;
```

调用 history.replace 同样会修改浏览器地址，但其如 browserHistory 的 replace 方法一样，仅替换某个历史记录：

```
const hashHisotry = createHashHistory();
console.log(location.href); // "https://example.com/a/b/c#/"
hashHisotry.push('/baz');
console.log(hashHisotry.length); // 1
hashHisotry.replace('/foo');
// 栈长度仍为 1
console.log(hashHisotry.length); // 1
```

在 history 库源码中，hashHistory 使用 window.location.replace 接口，以达到仅替换历史栈记录而不添加历史栈记录的目的：

```
// 以#号开头，仅替换 hash 部分
window.location.replace('#path');
```

注意，与 browserHistory 不同的是，hashHistory 在创建时没有 keyLength 选项，这对应到 hashHistory.push、hashHistory.repalce 方法不产生 key 值，即 hashHistory.location.key 为 undefined：

```
hashHisotry.replace('/foo');
console.log(hashHistory.location.key) // undefined
```

这里需要提一下 base 元素基准路径处理问题，如果 HTML 文档流中存在 base 元素，且 base 元素的 href 属性不为空，则在 history v4.10.0 中，替换 hash 的方法如下：

```
function replaceHashPath(path) {
  const hashIndex = window.location.href.indexOf('#');
  window.location.replace(
    window.location.href.slice(0, hashIndex >= 0 ? hashIndex : 0) + '#' + path
  );
}
```

history.replace 调用的 replaceHashPath 会在 window.location.href 没有#号时，得到 window.location.href.slice(0,0)的结果。由于 window.location.href.slice(0,0)返回空字符串，因此实际会调用：

```
window.location.replace("#some-path");
```

但是对于有 base 元素的 HTML 文档来说：

```
<base href="/base/"/>
```

window.location.replace 仅传入 hash 字符串的调用会将 window.location.pathname 改成 base 的值，原 pathname 将丢失，如：

```
console.log(location.href); // https://example.com/foo/baz
// 在调用 window.location.replace("#path") 后
window.location.replace("#path")
// window.location.pathname 将变为 base 元素的 href 属性值
console.log(location.href); // https://example.com/base/#path
```

假若当前路径为 https://example.com/foo/baz，且有 href 属性值为/base/的 base 元素存在，在调用 window.location.replace("#path")后，原路径中/foo/baz 将会丢失。若要避免丢失，则需要将 /foo/baz 也一起传入，如 window.location.replace("/foo/baz/#path")。

在 history v4.10.1 中，replaceHashPath 方法改为了：

```
function stripHash(url) {
  const hashIndex = url.indexOf('#');
  // 剥除 hash
  return hashIndex === -1 ? url : url.slice(0, hashIndex);
}
function replaceHashPath(path) {
  // 把 hash 剥除，保留非 hash 部分，再重新拼接
  window.location.replace(stripHash(window.location.href) + '#' + path);
}
```

在调用 window.location.replace 时会保留 pathname 部分，history v4.10.1 修复了这个问题，使用 history v4.10.1 以下版本的开发者需注意这个问题。

3. history.go

其实现与 browserHisotry 的 history.go 一致，调用全局的 window.history.go 方法：

```
function go(n) {
   globalHistory.go(n);
}
function goBack() {
  go(-1);
}
function goForward() {
```

```
    go(1);
  }
```

当调用 history.go 方法时，history 的状态更新由 hashchange 事件监听函数进行处理，具体内容将在 2.5.4 节介绍。

注意，在 Firefox 浏览器中，通过 widnow.history.go 移动指针而改变 hash 会使得页面重新刷新加载，这与 Chrome 等浏览器的行为不一致。

2.3.3 history 监听

hashHistory 的监听函数 listen 接口与 browserHistory 的接口完全一致：

```
const history = createHashHistory();
history.listen((location,action)=>{
    // location 为当前路由路径
    // action 为当前路由路径变化的行为类型，有 POP、PUSH 和 REPLACE 3 种
})
```

当 location 记录发生变化时，会触发 listen 监听的回调函数。回调函数参数列表中第一个参数 location 为变化后的地址信息；第二个参数 action 为跳转类型，分别为"POP""PUSH""REPLACE"。

history 的设计风格都是在调用订阅函数之后返回一个取消订阅的函数，可调用取消订阅函数取消 history 的监听：

```
const history = createHashHistory();
const unlisten = history.listen((location,action)=>{
   // ……
})
// 取消监听
unlisten()
```

类似于 addEventListener 可对单个事件添加多个回调函数，history.listen 同样可多次调用，添加多个监听函数：

```
const history = createHashHistory();
const unlisten1 = history.listen((location,action)=>{
   // ……
   console.log('location changed 1')
})
```

```
  const unlisten2 = history.listen((location,action)=>{
    // ……
    console.log('location changed 2')
})
history.push('/foo')
// 打印 location changed 1
// 打印 location changed 2
// 可分别调用 unlisten1、unlisten2 取消 location 变化的监听
```

对于多次调用监听函数，也会返回响应的注销方法，可分别调用注销方法取消 history 的监听。

对于 history.listen，类似于浏览器的事件订阅，当监听到 location 变化时，会通知监听函数。在 history.listen 的源码实现上，调用 history.listen 将事件回调函数保存到一个监听列表中。监听列表维护了一个回调函数数组，当需要取消响应事件时，则将需要取消的回调函数从这个回调函数数组中移除。当需要通知事件时，对回调函数数组中的每一个函数依次进行调用。其源码如下：

```
let listeners = [];
// 添加回调函数，返回取消订阅函数
function appendListener(fn) {
    let isActive = true;
    function listener(...args) {
      if (isActive) fn(...args);
    }
    // 保存事件回调函数到监听列表中
    listeners.push(listener);
    // 返回事件取消函数
    return () => {
      isActive = false;
      listeners = listeners.filter(item => item !== listener);
    };
}
// 通知所有回调函数
function notifyListeners(...args) {
    listeners.forEach(listener => listener(...args));
}
function listen(listener) {
    // 添加 listener 到监听列表中
    const unlisten = transitionManager.appendListener(listener);
    // checkDOMListeners(1)保证多次调用 listen 只会注册一次 popstate 事件或
```

```
  // hashchange 事件
  checkDOMListeners(1);
  // 返回事件取消函数
  return () => {
    checkDOMListeners(-1);
    unlisten();
  };
}
```

在上例源码中，listen 将通过 checkDOMListeners(1)监听原生的浏览器事件，如 popstate 事件，参数中的数字表示 listen 的调用计数。调用一次 listen，内存中变量会加 1；当移除事件时，内存中变量会减 1。只有当值为 0 时，调用 checkDOMListeners(1)才会注册一次相关原生事件，这样避免了调用 listen 多次注册浏览器事件。

2.3.4　history.createHref

createHref 方法可将 location 对象转换为对应的 URL 字符串，在 React Router 源码中使用到了该方法，如在 7.1.2 节中，Link 组件内部使用 createHref 方法创建 a 标签的 href 属性值。由于该方法为 history 对象的一个属性，所以开发者也可通过 history.createHref 使用该方法。

在使用 hashHistory.createHref 方法时，注意 createHref 不会对原字符串做任何编解码处理：

```
// 转换 location 对象为字符串
const encodedHref = history.createHref({
  pathname: '/%23abc'
});
// 不对字符串做编码处理
const unencodedHref = history.createHref({
  pathname: '/#abc'
});
console.log(encodedHref);    // #/%23abc
console.log(unencodedHref);  // '#/#abc'
```

并且 hashHistory 的 createHref 会判断 HTML 文档流中是否有 href 属性的 base 元素，其源码实现如下：

```
function createHref(location) {
  const baseTag = document.querySelector('base');
  let href = '';
```

```
  if (baseTag && baseTag.getAttribute('href')) {
    href = stripHash(window.location.href);
  }
  return href + '#' + encodePath(basename + createPath(location));
}
```

在文档流中没有 base 元素的情况下，调用 history.createHref 创建路径，得到的字符串将以 "#" 开头：

```
const loc = {
  pathname: "/the/path",
  search: "?the=query",
  hash: "#the-hash"
}
// 在没有 base 元素的情况下
history.createHref(loc)
// 得到: "#/the/path?the=query#the-hash"
```

如果文档流中有 base 元素且 href 有值 (任何值均可)，假设当前路径为/foo/baz，则调用 history.createHref 会得到：

```
// 当前路径为/foo/baz
const loc = {
  pathname: "/the/path",
  search: "?the=query",
  hash: "#the-hash"
}
// 在有 base 元素且 href 有值的情况下
history.createHref(loc)
// 得到: "/foo/baz#/the/path?the=query#the-hash"
```

此种情况将会得到包括浏览器 pathname 的完全的 href 路径。

由于在有 base 元素的情况下需要使用完全的路径设置到 a 标签的 href 属性上，才能使得浏览器识别出该 a 标签是否被访问过，因此，createHref 源码才需要判断文档流中的 base 元素并拼接出全路径，以便正确设置 a 标签的属性值。

2.4 memoryHistory

2.4.1 创建 memoryHistory

可通过调用 createMemoryHistory 创建 memoryHistory：

```
import { createMemoryHistory } from 'history';
const memoryHistory = createMemoryHistory()
```

对于内存路由，其运行环境通常不在浏览器内，一般作为测试使用或如 React Native 原生环境。在创建 memoryHistory 时，除了 history 配置，如 keyLength、getUserConfirmation，还可传入 initialEntries、initialIndex，其声明如下：

```
export interface MemoryHistoryBuildOptions {
  getUserConfirmation?: typeof getUserConfirmation;
  initialEntries?: string[];
  initialIndex?: number;
  keyLength?: number;
}
```

注意，basename 在 memoryHistory 中不被支持。

initialEntries 类似于 Browser Router 或 Hash Router 的历史栈，它确定了初始化的栈内容。由于是内存路由，这个历史栈仅能由 history 库进行记录。initialIndex 表示初始的栈指针位置。它们默认的值如下：

```
const {
  initialEntries = ['/'],
  initialIndex = 0,
  keyLength = 6
} = props;
```

initialEntries 的默认值为拥有初始入口 "/" 的一个栈记录，initialIndex 的默认值为 0。

对于 memoryHistory，其除了通用的 history 的属性，还多出 index、entries 和 canGo 属性。

```
export interface MemoryHistory<HistoryLocationState = LocationState>
extends History<HistoryLocationState> {
  index: number;
  entries: Location<HistoryLocationState>[];
  canGo(n: number): boolean;
}
```

entries 为历史栈数组，比起 browserHistory 与 hashHistory，memoryHistory 能获取所有的历史记录，如上一个导航地址、第一个导航地址等。index 为当前历史栈指针的指针位置，需要获取当前的地址，可从 history.entries[history.index] 中获取，其也等价于 history.location。history.length 即等价于 entries.length。对于 memoryHistory，其 canGo 属性用来判断跳转位置 n 是否可以跳转。

```
const history = createMemoryHistory();
// 读取整个历史栈
console.log(history.entries); // ["/"]
// 读取当前的栈指针位置
console.log(history.index) // 0
// 读取当前栈记录
console.log(history.entries[history.index]) // "/"
history.canGo(10) // false
```

2.4.2 history 导航

1. history.push

与 browserHistory 一样，其签名为：

```
push(path: string, state?: HistoryLocationState): void;
push(location: LocationDescriptorObject<HistoryLocationState>): void;
```

对于 memoryHistory，其 location 存储在内存中，在调用 history.push 后，得到的 location 既可通过 memoryHistory.location 获取，也可通过访问历史栈获取：

```
const memoryHistory = createMemoryHistory();
memoryHistory.push({
  pathname:'/baz',
  search:'?name=2',
  hash: "#c"
});
memoryHistory.push('/foo');
console.log(memoryHistory.location) // /foo
memoryHistory.entries[history.index] === memoryHistory.location //true
```

与 browserHistory、hashHistory 一样，memoryHistory 不仅支持调用 block 阻止跳转，还支持相对路径导航、保存 state 等：

```
const unBlock = memoryHistory.block();
```

```
// 此调用不产生任何效果
memoryHistory.push('/foo');
// 取消阻止跳转
unBlock();
// 相对路径导航
memoryHistory.push('./biz');
// 改变 hash
memoryHistory.push('#home');
memoryHistory.push('/baz',{a:1});
console.log(memoryHistory.location.state) // {a:1}
```

注意,在 memoryHistory 导航过程中,所有信息都将保存到 location 中,不像 browserHistory 的 key 与 state 一样存在于 window.history.state 中进行持久化存储,memoryHistory 导航过程中所有的 key 与 state 均存储在自身维护的内存中。

```
memoryHistory.push('/fooo',{a:1});
memoryHistory.push('/fooz',{a:2});
// 读取栈指针位置 1
memoryHistory.entries[1].key // gfwqew
memoryHistory.entries[1].state // {a:1}
//读取栈指针位置 2
memoryHistory.entries[2].key // bhrwwq
memoryHistory.entries[2].state // {a:2}
```

对于 history.push 的实现,其模拟了浏览器管理历史记录的方法:

```
// 当前栈指针位置
const prevIndex = history.index;
// 添加栈记录,指针移动+1
const nextIndex = prevIndex + 1;
// 复制一份
const nextEntries = history.entries.slice(0);
// 栈长度大于栈指针的情况,可查看 1.2.1 节
if (nextEntries.length > nextIndex) {
  // 更新历史栈
  nextEntries.splice(nextIndex, nextEntries.length - nextIndex, location);
} else {
  // 添加历史栈
  nextEntries.push(location);
}
```

2. history.replace

history.replace 可替换历史栈内容，与 browserHistory、hashHistory 等均一致，其签名如下：

```
replace(path: Path, state?: HistoryLocationState): void;
replace(location: LocationDescriptorObject<HistoryLocationState>): void;
```

history.replace 不增加栈记录：

```
const memoryHistory = createMemoryHistory();
memoryHistory.push('/baz');
console.log(memoryHistory.length); // 1
memoryHistory.replace('/foo');
// 栈长度仍为1
console.log(memoryHistory.length); // 1
```

对于 history.replace，其内部实现非常简单，使用新的 location 替换历史栈中对应位置的历史记录即可：

```
history.entries[history.index] = location;
```

3. history.go

memoryHistory 的 go 方法与 browserHistory、hashHistory 的 go 方法一样，都是移动栈指针，不改变栈的内容；但其在实现上与这两者不同的是，browserHistory 和 hashHistory 历史栈的栈指针由浏览器管理，而 memoryHistory 没有 history.go 方法可供调用以移动栈指针，所以 memoryHistory 自行维护栈指针，即便调用 history.go 方法也应自行控制栈指针。

```
function go(n) {
    // 计算要移动的栈指针位置
    const nextIndex = clamp(history.index + n, 0, history.entries.length - 1);
    const action = 'POP';
    // 获取栈信息
    const location = history.entries[nextIndex];
    // 跳转确认，将在2.5节介绍
    transitionManager.confirmTransitionTo(
      location,
      action,
      getUserConfirmation,
      ok => {
```

```
    if (ok) {
      setState({
        action,
        location,
        index: nextIndex
      });
    } else {
      // 在 block 时，通过 setState 触发一次更新，但是 location 不变
      setState();
    }
  }
);
}
```

history.go 方法对跳转距离 n 做了钳位控制：

```
const nextIndex = clamp(history.index + n, 0, history.entries.length - 1);
```

n 被限制在 0 与 history.entries.length-1 之间。

2.4.3　history 监听

与 browserHistory、hashHistory 一样，memoryHistory 也可对 location 变化进行监听：

```
const history = createMemoryHistory();
history.listen((location,action)=>{
    // location 为当前路由路径
    // action 为当前路由路径变化的行为类型，有 POP、PUSH 和 REPLACE 3 种
})
```

调用 history.push、history.replace、history.go 分别对应的 action 为 PUSH、REPLACE 和 POP。

由于内存路由没有外部的浏览器环境，不需要对任何的外部事件进行监听，所以 listen 源码实现可将回调函数存储到内存数组中即可：

```
function listen(listener) {
  return transitionManager.appendListener(listener);
}
```

2.5 history 库原理

2.5.1 history 库的运行流程

无论 browserHistory、hashHistory 还是 memoryHistory，其通用运行流程都如图 2-2 所示。

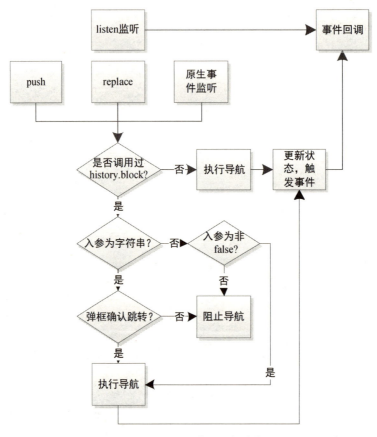

图 2-2　history 库的运行流程

history 库统一维护了回调函数数组，用于存放 listen 的响应函数。当调用 push、replace 方法时，会确认是否调用过 history.block。如果调用过且 history.block 入参不为 false，则在导航行为发生前会进行判定；如果入参为字符串，则会将字符串内容用于提示用户，由用户判定是否执行导航操作。如果 history.block 调用的入参为 false，则会强制阻止导航并且没有任何提示给用户。注意，在源码中仅判断值为 false 的入参，对于 history.block(true)、history.block(null) 等的调用，将不会有阻止导航的效果。当导航成功之后，history 在内部会收集所有的变动，如 location 的改变、history.length 的改变、action 的改变，并将所有的变动更新到 history 各属性中。在更新完 history 的状态以后，将触发 listen 函数的所有回调事件函数，将最新的变动 location 与 action 通知给各回调事件函数。注意，对于 memoryHistory，由于没有外部浏览器的运行环境，将没有浏览器原生事件监听并允许导航的部分，history.push、history.go 等流程与图 2-2 均一致。

2.5.2 history 模拟历史栈

无论 browserHistory、hashHistory 还是 memoryHistory，对于 push 与 replace 方法，其在内部实现上都模拟了浏览器的历史栈管理能力。

1. browserHistory

在代码内部，browserHistory 维护了 key 的数组 allKeys，用此数组记录历史栈的情况，其 length 长度与 browserHistory.length 的长度一致，且将 key 作为 window.history.state 的一个属性持久化存储在浏览器中。key 在每次调用 push 方法时，都会生成一个随机值，用于唯一标识当前产生的路径，可在创建 history 时配置 keyLength 属性以控制其长度。push 方法模拟了浏览器的历史栈管理能力：

```
// 获取之前的顶层 key 的下标
const prevIndex = allKeys.indexOf(history.location.key);
const nextKeys = allKeys.slice(0, prevIndex === -1 ? 0 : prevIndex + 1);
// 保证顶层是新加入的 location.key
nextKeys.push(location.key);
allKeys = nextKeys;
setState({ action, location });
```

allKeys 作为一个内部维护的数组，其记录了历史栈中栈记录的身份标识。其中 key 的生成如下：

```
function createKey() {
  return Math.random()
    .toString(36)
    .substr(2, keyLength);
}
```

与 replaceState 行为一致，在调用 replace 时也会替换对应栈记录位置的 key：

```
const prevIndex = allKeys.indexOf(history.location.key);
// 替换 key
if (prevIndex !== -1) allKeys[prevIndex] = location.key;
```

keyLength 使用频率较低，预计将在 history v5.x 之后的版本中移除。

如图 2-3 所示，browserHistory 虽然不能直接得知浏览器的历史栈，但可以从内存维护的 allKeys 数组中获得所有 key 的情况。

allKeys = ["gewr32","wrezvx","vdw35r"]

图 2-3　浏览器历史栈与 browserHistory 中的 allKeys

如从 window.hisotry.state.key 中获取到的值为 gewr32，则可从 allKeys 数组中判断出当前的栈指针位于栈底。由于在内存中维护了栈记录，在导航时能获得更加全面的信息；也可以阻止导航，这将在 2.5.5 节介绍。

2. hashHistory

hashHistory 没有持久化能力，其使用当前路径 URL 作为路径的唯一标识，而不是随机生成的 key，其维护的历史栈数组为 path 数组 allPaths。allPaths 的功能等同于 allKeys：

```
// 数组内的每一项标识每一次导航
allPaths = ["/a?name=1", "/b#foo", "/c"]
```

hashHistory 维护的历史栈数组 allPaths 同样会在调用 history.block 时发挥作用。

3. memoryHistory

与 browserHistory、hashHistory 不同的是，memoryHistory 没有外界的副作用。由于 memoryHistory 模仿了浏览器的管理能力，其内部自身应维护一个历史栈以便 history.go(-1)等方法能正常运行。与 browserHistory、hashHistory 不同，memoryHistory 维护历史栈的目的是提供类似浏览器对历史栈的管理，而不是作为 history.block 的参考历史栈。而且，memoryHistory 可通过 history.entries 接口获取整个历史栈信息，这是 browserHistory 与 hashHistory 不具备的。

2.5.3 browserHistory 事件处理

在 browserHistory 中，监听 popstate 事件用以获知浏览器地址的改变，如调用 history.go，则单击浏览器的"前进"和"后退"按钮，通过 window.location.hash 改变 hash 等都会触发 popstate 事件。在 popstate 事件中，browserHistory 会更新其维护的地址信息并通知 history 的监听者。注意，history.pushState、history.replaceState 并不会触发 popstate 事件。在源码中，browserHistory 在 needsHashChangeListener 为 true 时监听了 hashchange 事件，源码如下：

```
function checkDOMListeners(delta) {
  listenerCount += delta;
  if (listenerCount === 1 && delta === 1) {
    window.addEventListener(PopStateEvent, handlePopState);
    if (needsHashChangeListener)
      // 在某些场景中既监听了popstate事件，又监听了hashchange事件
      window.addEventListener(HashChangeEvent, handleHashChange);
  } else if (listenerCount === 0) {
    /* 取消事件监听 */
  }
}
```

在功能职责上，browserHistory 应该对 URL 的所有变化做出反应。相对于 Chrome 等浏览器 hash 变化会触发 popstate 事件，在 IE 浏览器上，改变 hash 不会触发 popstate 事件，而会触发 hashchange 事件。因此，在 IE 浏览器上，即便是 browserHistory，也需要对 hashchange 事件进行监听，以应对 URL 的所有变化。

2.5.4 hashHistory 事件处理

hashHistory 监听了浏览器的 hashchange 事件,用以获知浏览器地址的变化。对于 hashHistory,除了单击浏览器的"前进"和"后退"按钮时会触发 hashchange 事件,hashHistory 的 push 或 replace 方法由于改变了 hash,同样会触发 hashchange 事件,这与 browserHistory 在内部处理上会有一些不同。

由于 hashHistory 的 push 或 replace 方法改变了 hash 会触发 hashchange 事件,因此 hashchange 事件的事件响应函数需要判断事件的触发来源,是来源于栈指针移动,如 go(1)、go(-1),还是使用 push、replace 方法改变 hash 产生的 hashchange 事件。若不进行判断,则所有的导航都将在 hashchange 的事件响应函数中更新导航状态的 action 为"POP"。在源码实现上,在 push 或 replace 方法中会触发一次导航状态为"PUSH"或"REPLACE"的导航行为,且会对 push 或 replace 方法的导航路径做一次存储标记。如果在 hashchange 事件中获得的路径与使用 push 或 replace 方法设置的路径一致,则表明已经通过使用 push 或 replace 方法产生了导航行为,那么在 hashchange 事件中将不对此次事件响应做任何处理:

```
function handleHashChange() {
  const path = getHashPath();
  const encodedPath = encodePath(path);

  if (path !== encodedPath) {
    // 保证 hashType 的正确,如 hashbang 为#!/
    replaceHashPath(encodedPath);
  } else {
    const location = getDOMLocation();
    const prevLocation = history.location;
    // 如果前后两次 hashchange 事件触发得到的 location 一致,则不做处理
    if (!forceNextPop && locationsAreEqual(prevLocation, location)) return;
    // push 或 replace 方法中存储的路径,再次触发 handleHashChange 将被忽略
    if (ignorePath === createPath(location)) return;
    ignorePath = null;
    // 执行更新 location 操作
    handlePop(location);
  }
}
```

在 browserHistory 中，browserHistory 的 popstate 监听函数不会特意对两次相同导航做出处理，这意味着同样的路径变化都会触发 history.listen 监听的回调函数。如前后两次路径相同的回退操作：

```
browserHistory.go(-1);
// location.pathname 为 /foo
// browserHistory.listen 监听函数触发
browserHistory.go(-1);
// location.pathname 为 /foo
// browserHistory.listen 监听函数触发
```

而如果 hashHistory 在某些情况下进行两次路径相同的回退操作，则回调函数不会触发两次：

```
const hashHistory = createHashHistory();
hashHistory.listen(() => {
  console.log("hashHisotry change");
});
  hashHistory.go(-1);
  // location.pathname 为 /foo；hashHistory.listen 监听函数触发
  // 打印 hashHisotry change
  hashHistory.go(-1);
  // location.pathname 不改变，为/foo；hashHistory.listen 监听函数不会触发
  // 不打印 hashHisotry change
```

这也是模拟了原生 hashchange 事件的行为：

```
window.addEventListener("hashchange", () => {
  console.log("原生 hash change");
});
location.hash = "#/three"
// 打印:原生 hash change
location.hash = "#/three"
// 不打印任何信息，历史栈不变
```

但与原生行为不同的是，进行 2 次路径相同的 push 操作或 replace 操作都会触发 history.listen 监听的回调函数。但是，相同路径 push 函数调用会得到对应的警告，为了与原生行为保持一致，其不会使历史栈增加：

```
const hashHistory = createHashHistory();
hashHistory.listen(() => {
  console.log("hashHisotry change");
});
```

```
hashHistory.push('/a')
// 打印：hashHisotry change
hashHistory.push('/a')
// 打印：hashHisotry change
// warning Hash history cannot PUSH the same path; a new entry will not
be added to the history stack
```

而 replace 函数调用不会得到警告，因为其不会影响历史栈的记录数量。

2.5.5　history.block 原理解析

history.block 对应 history v3.x 中的 listenBefore 监听方法，history.block 同时也服务于 React Router 的 Prompt 组件。在 2.1 节曾介绍过，history.block 的目的是在一些表单还未输入完成的页面，对用户的意图跳转行为进行阻止或弹窗确认。

那么 history.block 是如何"拦截"浏览器导航的呢？以 browserHistory 为例，对于 history.push 和 history.replace 方法，其会在真正跳转前调用 transitionManager.confirmTransitionTo 进行跳转确认，各类 history 都会在内部创建 transitionManager：

```
const transitionManager = createTransitionManager();
```

transitionManager.confirmTransitionTo 的主要作用是确认是否应该阻止跳转，其实现如下：

```
function confirmTransitionTo(
    location,
    action,
    getUserConfirmation,
    callback
) {
    if (prompt != null) {
        // 若 prompt 不为 null，就会执行跳转确认检查
        const result =
            typeof prompt === 'function' ? prompt(location, action) : prompt;
        if (typeof result === 'string') {
            if (typeof getUserConfirmation === 'function') {
                // 用户确认
                getUserConfirmation(result, callback);
            } else {
                // 允许跳转
                callback(true);
```

```
        }
      } else {
        // 为 false 时不允许跳转
        callback(result !== false);
      }
    } else {
      // 允许跳转
      callback(true);
    }
  }
```

而 history.block 的真正作用是设置 confirmTransitionTo 中的 prompt 函数：

```
let prompt = null;
  function setPrompt(nextPrompt) {
    prompt = nextPrompt;
    return () => {
      if (prompt === nextPrompt) prompt = null;
    };
  }
  function block(prompt = false) {
    // 调用 block 时设置 prompt 函数
    const unblock = transitionManager.setPrompt(prompt);
    if (!isBlocked) {
      checkDOMListeners(1);
      isBlocked = true;
    }
    return () => {
      if (isBlocked) {
        isBlocked = false;
        checkDOMListeners(-1);
      }
      return unblock();
    };
  }
```

所以，在执行 block 后，prompt 变量不为 null。这样在使用 history 每次调用 history.push、history.replace 时，将检查 prompt，因为 prompt 不为 null 将进行跳转拦截。但如果调用 history.go

或者单击浏览器的"前进"或"后退"按钮跳转该如何"拦截"？事实上，单击"前进"或"后退"按钮只能监听到 popstate、hashchange 事件，但是此刻浏览器地址已经改变了，所以为了"拦截"跳转，要做的是在地址发生变化后进行人工恢复。在此情况下，可调用 revertPop 进行地址恢复：

```
// fromLocation 为浏览器变化后已经发生跳转的地址
function revertPop(fromLocation) {
  // 希望恢复的地址
  const toLocation = history.location;
  // allKeys 维护了跳转历史栈
  let toIndex = allKeys.indexOf(toLocation.key);
  if (toIndex === -1) toIndex = 0;
  let fromIndex = allKeys.indexOf(fromLocation.key);
  if (fromIndex === -1) fromIndex = 0;
  // 计算恢复地址所需的移动距离
  const delta = toIndex - fromIndex;
  if (delta) {
    // 设置标志位,当 go 再次触发 popstate 事件后,不用再调用 transitionManager.confirmTransitionTo 进行跳转确认
    forceNextPop = true;
    // 执行跳转恢复
    // 再次触发 popstate 事件，根据 delta 值进行恢复
    go(delta);
  }
}
```

地址恢复的依据是 2.5.2 节介绍过的内存栈。根据内存栈（如 browserHistory 的 allKeys）、已经发生跳转的地址，以及希望恢复的地址三部分信息，可以计算出恢复到目标地址所需的跳转恢复变量 delta 值。

在 browser 路径下，进行人工恢复会触发两次 popstate 事件，第一次在单击浏览器的"后退"或"前进"按钮时触发，第二次在调用 history.go 方法时触发。第一次触发 popstate 事件不会通知 history.listen 监听器，只有第二次触发 popstate 事件后在使用 history.go 恢复地址时，才会重新触发一次 history.listen 监听器，并触发一次渲染。看起来路径是不变的，原路径对应的组件会接受一次 props 更新。

2.6　history 库限制

2.6.1　history.block 的使用限制

值得一提的是，history.block 存在以下几个问题。

1．刷新页面后历史栈丢失

对于大多数情况，history.block 都能很好地工作，但是如果刷新页面，则原内存将丢失。以 browserHistory 为例，allKeys 会重新初始化为空，原历史栈记录将全部清空，而浏览器自身的历史栈记录在刷新后是不会丢失的，这种跟浏览器行为不一致之处造成了问题。例如对于 browserHistory，有/a、/b、/c 页面，从/a 页面开始，按如下路径进行跳转：/a→/b→/c；如果在/c 页面进行了 history.block 阻止，这时由于每次导航都在内存中维护了历史栈记录，知道整个历史栈的情况，所以能在调用 history.go(-1)后退一次后，browserHistory 再重新调用 history.go(1)前进一次，即一次"后退"再"前进"，进而恢复到/c，达到了路径不变的效果。如果在/c 页面调用 history.block 后再刷新，即在内存中的历史栈内容丢失之后，先调用 history.block，再调用 history.go(-1)或 history.go(-2)，或者单击浏览器地址栏侧的"后退"按钮。这时由于内存中历史栈被清空，没有可供参考的回退历史记录，因此在内部计算得到的跳转恢复变量 delta 值为 0，这时虽然 history.go(-1)或 history.go(-2)已经生效，URL 已经改变为/b 或/a，但是 URL 地址并未如期恢复到/c，最终出现 URL 与页面内容不一致的情况。

2．移动端的限制

如果遇到移动端历史栈记录数量的上限为 100 或者在移动端 window.confirm 不生效等问题，则需要 Native 开发者配合支持。比如在 iOS 12 移动端场景下，history.block 使用的默认 prompt 为 window.confirm，其不会如 PC 页面一样有弹窗提示，而是默认返回"false"，这将阻止浏览器跳转，而不是弹出提示框让用户进行确认。

3．hash 路由问题

如果是 hash 路由，则有两个问题，一个问题是使用 path 作为历史栈记录，path 路径与

browserHistory 的 key 作用一样，但没有使用 key 的原因是考虑到 hash 路由没有 window.history.state 可以使用。所以该问题是路径重复的问题，如栈中有两个 /b 路径，在寻找 /b 路径时，将不知道以哪一个为准。例如当前栈路径为 /a→/b→/c→/b，如果此时在第一个 /b 页面调用了 history.block，再单击浏览器的"后退"按钮，强制返回到了 /a，则要恢复到 /b，这时将不知道恢复到哪个 /b。history 将会错误地恢复到最后一个 /b 页面，与预期不符。

另一个问题是可以在地址栏中手动更改 hash 进行一次入栈操作，而事实上由于手动更改 hash 没有调用 push 方法，内存中的历史栈不会感知到手动更改 hash 导致的入栈。若当前路径为 /a→/b→/c，且在 /c 处调用了 history.block，如果此时在地址栏中输入 /b，则路径将为 /a→/b→/c→/b，这时应该调用 history.go(-1) 恢复到 /c，以达到阻止跳转的目的。由于内存中的栈只有 /a→/b→/c，history 将错误地认为当前的 /b（手动输入的 /b）为 /a 之后的 /b，则要从 /b 恢复到 /c，应该是从第二个 /b 前进一步到 /c，因而会错误地调用 history.go(1) 进行恢复，而不是调用 history.go(-1)。结果是路径已经变为了 /b（history 判断需要前进一步，但是此时已经是最后的位置），但是页面还是 /c 的页面（页面被 history.block 所阻止）。

事实上，上述问题的本质在于 history 库无能力真正"阻止" URL 的改变，目前 history v4.x 将历史栈保存在内存中，从而造成了种种问题。

注意，对于 memoryHistory.block 方法是可以放心使用的，因为路径都在内存中进行了维护，可认为 memoryHistory 没有浏览器这个"副作用"。由于其没有浏览器的"后退"按钮和对应的浏览器 popstate 等事件，所有的前进、后退动作都在内存中操作，所以如果进行了 history.block 操作，则无须进行如 browserHistory、hashHistory 中的地址恢复行为，仅需执行一次更新操作即可。

2.6.2 decodeURI 解码问题

在 push、replace 方法或者浏览器"前进"和"后退"时的事件回调函数中，history 源码都会调用 createLocation 创建 location 对象。

```
export function createLocation(path, state, key, currentLocation) {
  /*
    ......
  创建 location 对象
  */
  try {
    // 调用 decodeURI 解码 pathname
```

```
      location.pathname = decodeURI(location.pathname);
    } catch (e) {
      if (e instanceof URIError) {
        throw new URIError(
          'Pathname "' +
            location.pathname +
            '" could not be decoded. ' +
            'This is likely caused by an invalid percent-encoding.'
        );
      } else {
        throw e;
      }
    }
    /*
      ...
处理 location
    */
    return location;
}
```

每次进行相关操作都会对 pathname 解码一次,但是并没有对 search 与 hash 进行相关的操作。

1. search 没有解码带来的问题

以 browserHistory 为例,由于 history.push/replace 方法最终会调用浏览器的 pushState/replaceState 方法,因此如果 pushState/replaceState 方法入参中有 Unicode 字符,则浏览器会将非特殊的 Unicode 字符的 UTF-8 码中的每个字节加上 "%" 作为实际地址,window.location.search 或者 window.location.hash 得到的路径中都将带有 "%"。但是在 history 库中,如果调用 history 库的方法,如 history.push/replace,则参数列表中传入的字符串也将同步到 history.location.search 或者 history.location.hash 中,将不会做编码处理:

```
browserHistory.push('/one/two?name=中文#英文')
// 使用 history 库的方法不对 search 与 hash 做编码处理,即不做任何处理
console.log(browserHistory.location.search) // "?name=中文"
console.log(browserHistory.location.hash) // "#英文"
// 但是若从浏览器 window.location 中获取 search 与 hash
console.log(window.location.search); // "?name=%E4%B8%AD%E6%96%87"
console.log(window.location.hash); // "#%E8%8B%B1%E6%96%87"
```

```
// decodeURIComponent('?name=%E4%B8%AD%E6%96%87') === '?name=中文'
// decodeURIComponent('#%E8%8B%B1%E6%96%87') === '#英文'
```

从浏览器记录的角度来看，其记录 URL 字符串为编码后的字符串。此时，如果进行浏览器的"前进"和"后退"操作，则 browserHistory 将更新地址，browserHistory 的 search 和 hash 此时将从 window.location 中获取，获取到的值将为编码后的带"%"的字符串，因此造成 search 与 hash 前后不一致的问题：

```
browserHistory.push('/one/two?name=中文#英文')
// 使用 push 等方法，字符串不变化
console.log(browserHistory.location.search) // "?name=中文"
console.log(browserHistory.location.hash) // "#英文"
// 在浏览器中进行一次"后退"操作，再进行一次"前进"操作，重新回到当前路径，但是 location
// 由 window.location 提供
console.log(browserHistory.location.search) // "?name=%E4%B8%AD%E6%96%87"
console.log(browserHistory.location.hash) // "#%E8%8B%B1%E6%96%87"
```

这在编码时需要注意。

2. pathname 解码

对于 pathname，在 history 源码中，设置 history.location 前对 pathname 做了一次 decodeURI 解码处理。因此解决了上述问题，无论是执行 push、replace 操作，还是浏览器在"前进"和"后退"时，如 browserHistory.location.pathname，都将得到解码后的字符：

```
browserHistory.push('/中文')
// 通过 push 等方法
// 内部执行一次 browserHistory.location.pathname = decodeURI('/中文')
console.log(browserHistory.location.pathname); // "/中文"
// 浏览器的行为，会对 Unicode 字符进行编码
console.log(window.location.pathname); // "/%E4%B8%AD%E6%96%87"
browserHistory.go(-1)
browserHistory.go(1)
// 在浏览器中进行一次"后退"操作，再进行一次"前进"操作，重新回到当前路径，但是 location
// 由 window.location 提供，浏览器提供的 pathname 为"/%E4%B8%AD%E6%96%87"
// 源码内部同样会执行
// browserHistory.location.pathname = decodeURI('/%E4%B8%AD%E6%96%87')
// 依然为传入的中文字符
console.log(browserHistory.location.pathname); // "/中文"
```

这解决了在导航过程中 pathname 前后不一致的问题，给路由匹配带来了两个好处。

1）编码便利

由于 browserHistory.location.pathname 前后提供的 pathname 一致，因此对于 Route 的 path 属性中的特殊字符，可不用进行编码（Route 将在第 6 章进行介绍）：

```
<Route path='/中文' />
```

提供给该 Route 的进行命中匹配的路径将永远为解码后的字符串，即不会提供 /%E4%B8%AD%E6%96%87 给该 Route 进行匹配。

2）参数解析

对于命名路由（命名路由将在第 6 章进行介绍），如：

```
<Route path='/foo/:name' />
```

当调用如 browserHistory.push('/foo/中文')时，由于得到的 browserHistory.location.pathname 都为解码后的字符串，因此获取 name 这个命名变量的值也会得到解码后的字符串，即：

```
match === {
  ...,
  params: { name: '中文' }
}
```

永远不会出现{name: "/%E4%B8%AD%E6%96%87"}这样的情况。

由于在源码中进行了 decodeURI 解码，因此开发者在编码过程中如声明路由、进行导航等操作时不需要关心字符的解码问题。

3. pathname 解码带来的问题

虽然前后一致的pathname带来了编码的方便,但是引入pathname解码同时也带来了一些问题。

前面对 push 等方法的调用，传入参数字符串中没有特殊字符 "%"。对于没有 "%" 的字符串的导航路径，则不会有相关问题。但是如果导航路径中有 "%"，则需要注意，由于浏览器使用百分号编码，如"/%E4%B8%AD%E6%96%87"，在每个字节前都加上了 "%"，因此此时的 "%" 被认为有特殊作用，后面跟一个字节的十六进制形式。如果希望导航路径中的 "%" 没有特殊作用，则需要对 "%" 进行一次编码，调用 encodeURI('%')得到 "%25"。下面以一个例子来说明。

如果不编码而直接调用，如：

```
browserHistory.push('/abc%d')
history.js:110 Uncaught URIError: Pathname "/abc%d" could not be decoded. This is likely caused by an invalid percent-encoding.
```

```
    at createLocation (history.js:110)
    at Object.push (history.js:372)
    at <anonymous>:1:16
```

对于 history 库来说，由于其在导航操作时会对 pathname 进行 decodeURI 解码操作，这就默认 "%" 为特殊符号。上述错误即产生于：

```
decodeURI('/abc%d')
VM3019:1 Uncaught URIError: URI malformed
    at decodeURI (<anonymous>)
    at <anonymous>:1:1
```

无法解码非法的 "%"，如果不希望此错误产生，能导航到/abc%d 路径，则可对路径部分进行一次 encode 操作：

```
browserHistory.push(encodeURI('/abc%d'))
// encodeURI('/abc%d') === "/abc%25d"
// 内部得到原字符串
// browserHistory.location.pathname = decodeURI(encodeURI('/abc%d'))
// 浏览器路径为
// console.log(window.location.pathname); // "/abc%d"
```

由于 history 库内部会执行一次 decodeURI 操作，因此得到的 browserHistory.location.pathname，即原路径/abc%d，不会出现报错。细心的读者会发现浏览器路径此时为/abc%d：

```
// console.log(window.location.pathname); // "/abc%d"
```

浏览器认为 "%" 有特殊意义，不会对其进行处理，上述操作等同于：

```
browserHistory.push(encodeURI('/abc%d'))
// 等同于
window.history.pushState(null,null,'/abc%d')
// console.log(window.location.pathname); // "/abc%d"
// 浏览器认为 "%" 有特殊意义，不会将其处理成 "%25"
```

由于当前路径为/abc%d，因此虽然编码过一次没有报错，但是如果在浏览器中执行一次后退操作，再执行一次前进操作，又回到此路径时，则会出现解码错误：

```
history.js:110 Uncaught URIError: Pathname "/abc%d" could not be decoded. This is likely caused by an invalid percent-encoding.
    at createLocation (history.js:110)
    at getDOMLocation (history.js:298)
    at handlePopState (history.js:317)
createLocation @ history.js:110
```

```
getDOMLocation @ history.js:298
handlePopState @ history.js:317
```

由于 window.location.pathname 为/abc%d，在导航过程中，当路径/abc%d 传入 history 库中后，同样会执行一次 decodeURI 操作，这时无法解析/abc%d 中非法的"%"字符，因而报错。

这就是在 history 库中引入 decodeURI 所带来的问题。

要解决此问题，需要使浏览器地址中无特殊意义的"%"也得到编码，即要得到：

```
// console.log(window.location.pathname); // "/abc%25d" %编码为%25
```

则调用 pushState 时需要传入对"%"进行编码后的字符串：

```
window.history.pushState(null,null,'/abc%25d')
```

由于 history 库会执行一次 decodeURI 操作，因此要想获得上述 pushState 的效果，则需要在调用 push/replace 方法时进行两次编码：

```
browserHistory.push(encodeURI(encodeURI('/abc%d')))
// encodeURI('/abc%d') === "/abc%25d"
// encodeURI("/abc%25d") === "/abc%2525d"
// 内部得到一次编码后的字符串/abc%25d
// browserHistory.location.pathname=decodeURI('/abc%2525d')='/abc%25d'
// 浏览器路径为
// console.log(window.location.pathname); // "/abc%25d"
```

第一次编码是为了对特殊字符进行编码，第二次编码是为了抵消 history 库中的 decodeURI 操作。当执行完两次编码操作后，调用 push 方法得到：

```
browserHistory.push(encodeURI(encodeURI('/abc%d')));
// 为编码后的值
// console.log(browserHistory.location.pathname); // "/abc%25d"
// console.log(window.location.pathname); // "/abc%25d"
```

注意，由于此时的 browserHistory.location.pathname 为编码后的值/abc%25d，因此获取原始值/abc%d 需要进行一次解码（decodeURI(browserHistory.location.pathname)）才可实现。

在这样处理之后，在浏览器中执行一次前进、后退操作，由于 window.location.pathname 为/abc%25d，因此在传入 history 库中后，要进行一次解码：

```
// 在浏览器前进、后退导航时，由 window.location.pathname 提供路径，而不用 push
//等方法提供
// 此时的 pathname 值为
browserHistory.location.pathname = decodeURI("/abc%25d") = "/abc%d"
```

这样最终解决了两处报错的问题，但是执行 push/replace 操作得到的 location 路径将与浏览器前进、后退时得到的 location 路径不一致：

```
// 对路径进行两次编码，在调用 browserHistory.push 后得到的编码路径
browserHistory.location.pathname = "/abc%25d"
// 浏览器前进、后退时得到的解码路径
browserHistory.location.pathname = "/abc%d"
```

由于有此种问题存在，在此场景中，使用 browserHistory.location.pathname 变量获取原始路径可能有编码和未编码两种情况，对此可引入一些帮助函数（如 safeDecode）来处理：

```
const safeDecodeURI = (pathname) => {
  try {
    return decodeURI(pathname);
  } catch (_error) {
    return pathname;
  }
}
safeDecodeURI(browserHistory.location.pathname)
// 无论是编码情况还是未编码情况，调用 safeDecodeURI 后得到的路径都为"/abc%d"
```

同时，由于获取路径存在两种情况，因此如在 Route 声明时，也需声明两条路径：

```
<Route path={["/abc%d","/abc%25d"]} />
```

这样的声明使得两条路径中的任意一条都能匹配成功。

正是由于存在上述问题，在 history 计划的 5.x 版本中，可能会将内部的 decodeURI 解码逻辑移除。但是如果移除了 decodeURI 解码，则编码的便利性也将消失，如对于"/中文"路径，Route 可能会接收到"/%E4%B8%AD%E6%96%87""/中文"两类字符串，这可能需要 React Router 库内部调用 safaDecode 等逻辑进行联动修改，才能兼容 history 的升级改动。

2.7 使用 history 替换页面 search 和 hash 示例

声明 replaceSearch、replaceHash 方法仅替换页面 search 和 hash 部分：

```
import * as queryString from "query-string";
import {History} from "history";
function replaceLocationDescriptor(history: History, option: replaceOption) {
  return history.replace({
    // 保留原有的地址信息
```

```
    ...history.location,
    ...option
  });
}
function replaceHash(history: History, hash: string) {
  // 仅传入替换的 hash
  return replaceLocationDescriptor(history, { hash });
}
function replaceSearch(history: History, search: string | Object) {
  let searchStr = search;
  if (typeof search === "object") {
    searchStr = queryString.stringify(search);
  }
  // 仅传入替换的 search
  return replaceLocationDescriptor(history, { search: searchStr as string });
}
```

通过仅替换 query 和 hash，保持 pathname 部分不变，达到仅改变 search 和 hash 的目的。

2.8　小结

本章介绍了 history 库中各历史对象的特性及基本使用方法，并介绍了 history 对象的底层原理、history 库的运行流程，以及 history 库的一些现有限制，全面总结了 history 库。同时，history 库作为 React Router 中的路径管理者，在 React Router 中发挥着非常重要的作用。在后面的章节中即将看到，React Router 对地址的响应，使用 React Router 进行导航，都将使用到 history 库，因此理解 history 库对掌握 React Router 有着非常重要的意义。

参考文献

https://github.com/ReactTraining/history.

第 3 章
React 相关知识

鉴于读者对 React 有一定的认识，且本书所有案例均使用 React Hooks 编写，以及在 React Router 源码中使用了 Context 等 React 特性，因此本章仅对 React 的 Context、Hooks 等部分特性进行介绍。对于其他 React 相关特性，读者可查阅相关资料进行学习。

3.1 Context

在 React 中，父组件通常将数据作为 props 传递给子组件。如果需要跨层级传递数据，那么使用 props 逐层传递数据将会使开发变得复杂。同时，在实际开发中，许多组件需要一些相同的东西，如国际化语言配置、应用的主题色等，在开发组件的过程中也不希望逐级传递这些配置信息。

在这种情况下，可以使用 React 的 Context 特性。Context 被翻译为上下文，如同字面意思，其包含了跨越当前层级的信息。

Context 在许多组件或者开发库中有着广泛的应用，如 react-redux 使用 Context 作为 Provider，提供全局的 store，以及 React Router 通过 Context 提供路由状态。掌握 Context 将会对理解 React Router 起到极大的帮助作用。这里以图 3-1 来说明 Context 如何跨组件传递数据。

在图 3-1 中，左侧组件树使用了逐层传递 props 的方式来传递数据，即使组件 B、组件 C 不需要关心某个数据项，也被迫需要将该数据项作为 props 传递给子组件。而使用 Context 来实现组件间跨层级的数据传递，数据可直接从组件 A 传递到组件 D 中。

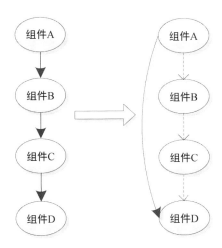

图 3-1　Context 跨组件传递数据

在 React v16.3 及以上版本中，可使用 React.createContext 接口创建 Context 容器。基于生产者-消费者模式，创建容器后可使用容器提供方（一般称为 Provider）提供某跨层级数据，同时使用容器消费方（一般称为 Consumer）消费容器提供方所提供的数据。示例如下：

```
// 传入 defaultValue
// 如果 Consumer 没有对应的 Provider，则 Consumer 所获得的值为传入的 1
const CountContext = React.createContext(1);
class App extends React.Component {
  state = {
    count: 0
  };
  render() {
    console.log("app render");
    return (
      <CountContext.Provider value={this.state.count}>
        <Toolbar />
        <button
          onClick={() =>
            this.setState(state => ({
              count: state.count + 1
            }))
          }
        >
          更新
```

```
      </button>
    </CountContext.Provider>
  );
 }
}
```

通过 setState 改变 count 的值,触发 render 渲染,Context.Provider 会将最新的 value 值传递给所有的 Context.Consumer。

```
class Toolbar extends React.Component {
  render() {
    console.log("Toolbar render");
    return (
      <div>
        <Button />
      </div>
    );
  }
}
class Button extends React.Component {
  render() {
    console.log("Button outer render");
    return (
      // 使用 Consumer 跨组件消费数据
      <CountContext.Consumer>
        {count => {
          // 在 Consumer 中,受到 Provider 提供数据的影响
          console.log("Button render");
          return <div>{count}</div>;
        }}
      </CountContext.Consumer>
    );
  }
}
```

在上例中,顶层组件 App 使用 CountContext.Provider 将 this.state.count 的值提供给后代组件。App 的子组件 Toolbar 不消费 Provider 所提供的数据,Toolbar 的子组件 Button 使用 CountContext.Consumer 获得 App 所提供的数据 count。中间层的 Toolbar 组件对数据跨层级传递没有任何感知。在单击"更新"按钮触发数据传递时,Toolbar 中的"Toolbar render"信息不会

被打印。每次单击"更新"按钮时，仅会打印"app render"与"Button render"，这是因为在 Provider 所提供的值改变时，仅 Consumer 会渲染，所以 Toolbar 中的"Toolbar render"不会被打印。

如果在 Toolbar 中也使用 Provider 提供数据，如提供的 value 为 500：

```
class Toolbar extends React.Component {
  render() {
    console.log("Toolbar render");
    return (
      <CountContext.Provider value={500}>
        <Button />
      </CountContext.Provider>
    );
  }
}
```

则 Button 中的 Consumer 得到的值将为 500。原因在于当有多个 Provider 时，Consumer 将消费组件树中最近一级的 Provider 所提供的值。这作为 React 的一个重要特性，在 React Router 源码中被大量应用。

注意，如果不设置 Context.Provider 的 value，或者传入 undefined，则 Consumer 并不会获得创建 Context 时的 defaultValue 数据。创建 Context 时的 defaultValue 数据主要提供给没有匹配到 Provider 的 Consumer，如果去掉 App 中的 Provider，则 Consumer 所获得的值为 1。

如果希望使用 this.context 方式获取 Provider 所提供的值，则可声明类的静态属性 contextType（React v16.6.0）。contextType 的值为创建的 Context，如：

```
const MyContext = React.createContext();
class MyClass extends React.Component {
  static contextType = MyContext;
  render() {
    // 获取 Context 的值
    let value = this.context;
  }
}
```

在 React v16.3 以前，不支持通过 createContext 的方式创建上下文，可使用社区的 polyfill 方案，如 create-react-context 等。

注意，组件的优化方法如 shouldComponentUpdate 或者 React.memo 不能影响 Context 值的传递。若在 Button 中引入 shouldComponentUpdate，则会阻止 Button 更新：

```
shouldComponentUpdate() {
    // 返回 false 阻止了 Button 组件的渲染，但是 Provider 提供的数据依然会提供到
    //Consumer 中
    // 不受此影响
    return false;
}
```

改变 Provider 所提供的值后，依然会触发 Consumer 的重新渲染，结果与未引入 shouldComponentUpdate 时一致。

3.2 Hooks

React Hooks 是 React v16.8 正式引入的特性，旨在解决与状态有关的逻辑重用和共享等问题。

在 React Hooks 诞生前，随着业务的迭代，在组件的生命周期函数中，充斥着各种互不相关的逻辑。通常的解决办法是使用 Render Props 动态渲染所需的部分，或者使用高阶组件提供公共逻辑以解耦各组件间的逻辑关联。但是，无论是哪一种方法，都会造成组件数量增多、组件树结构修改或者组件嵌套层数过多的问题。在 Hooks 诞生后，它将原本分散在各个生命周期函数中处理同一业务的逻辑封装到了一起，使其更具移植性和可复用性。使用 Hooks 不仅使得在组件之间复用状态逻辑更加容易，也让复杂组件更易于阅读和理解；并且由于没有类组件的大量 polyfill 代码，仅需要函数组件就可运行，Hooks 将用更少的代码实现同样的效果。

React 提供了大量的 Hooks 函数支持，如提供组件状态支持的 useState、提供副作用支持的 useEffect，以及提供上下文支持的 useContext 等。

在使用 React Hooks 时，需要遵守以下准则及特性要求。

- 只在顶层使用 Hooks。不要在循环、条件或嵌套函数中调用 Hooks，确保总是在 React 函数组件的顶层调用它们。
- 不要在普通的 JavaScript 函数中调用 Hooks。仅在 React 的函数组件中调用 Hooks，以及在自定义 Hook 中调用其他 Hooks。

3.2.1 useState

useState 类似于 React 类组件中的 state 和 setState，可维护和修改当前组件的状态。

useState 是 React 自带的一个 Hook 函数，使用 useState 可声明内部状态变量。useState 接收的参数为状态初始值或状态初始化方法，它返回一个数组。数组的第一项是当前状态值，每次渲染其状态值可能都会不同；第二项是可改变对应状态值的 set 函数，在 useState 初始化后该函数不会变化。useState 的类型为：

```
function useState<S>(initialState: S | (() => S)): [S, Dispatch <SetStateAction<S>>];
```

initialState 仅在组件初始化时生效，后续的渲染将忽略 initialState：

```
const [inputValue, setValue] = useState("react");
const [react, setReact] = useState(inputValue);
```

如上例中的 inputValue，当初始值传入另一个状态并初始化后，另一个状态函数将不再依赖 inputValue 的值。

使用 Hooks 的方式非常简单，引入后在函数组件中使用：

```
import { useState } from 'react';
function Example() {
  const [count, setCount] = useState(0);
  return (
    <div>
      <p>您点击了 {count} 次</p>
      <button onClick={() => setCount(count + 1)}>
      // 单击触发更新
      </button>
    </div>
  );
}
```

类似于 setState，单击按钮时调用 setCount 更新了状态值 count。当调用 setCount 后，组件会重新渲染，count 的值会得到更新。

当传入初始状态为函数时，其仅执行一次，类似于类组件中的构造函数：

```
const [count, setCount] = useState(()=>{
    // 可执行初始化逻辑
    return 0
});
```

此外，useState 返回的更新函数也可使用函数式更新：

```
setCount(preCount => preCount + 1)
```

如果新的 state 需要依赖先前的 state 计算得出，那么可以将回调函数当作参数传递给 setState。该回调函数将接收先前的 state，并将返回的值作为新的 state 进行更新。

注意，React 规定 Hooks 需写在函数的最外层，不能写在 if…else 等条件语句中，以此来确保 Hooks 的执行顺序一致。

3.2.2　useEffect

1. 副作用

在计算机科学中，如果某些操作、函数或表达式在其局部环境之外修改了一些状态变量值，则称其具有副作用（side effect）。副作用可以是一个与第三方通信的网络请求，或者是外部变量的修改，或者是调用具有副作用的任何其他函数。副作用并无好坏之分，其存在可能影响其他环境的使用，开发者需要做的是正确处理副作用，使得副作用操作与程序的其余部分隔离，这将使得整个软件系统易于扩展、重构、调试、测试和维护。在大多数前端框架中，也鼓励开发者在单独的、松耦合的模块中管理副作用和组件渲染。

对于函数来说，无副作用执行的函数称为纯函数，它们接收参数，并返回值。纯函数是确定性的，意味着在给定输入的情况下，它们总是返回相同的输出。但这并不意味着所有非纯函数都具有副作用，如在函数内生成随机值会使纯函数变为非纯函数，但不具有副作用。

React 是关于纯函数的，它要求 render 纯净。若 render 不纯净，则会影响其他组件，影响渲染。但在浏览器中，副作用无处不在，如果希望在 React 中处理副作用，则可使用 useEffect。useEffect，顾名思义，就是执行有副作用的操作，其声明如下：

```
useEffect(effect: React.EffectCallback, inputs?: ReadonlyArray<any> | undefined)
```

函数的第一个参数为副作用函数，第二个参数为执行副作用的依赖数组，这将在下面的内容中介绍。

示例如下：

```
const App = () => {
  const [value, setValue] = React.useState(0);
  // 引入 useEffect
  React.useEffect(function useEffectCallBack() {
    // 可执行副作用
    // 在此进行数据请求、订阅事件或手动更改 DOM 等操作
    const nvDom = document.getElementById("content");
```

```
      console.log("color effect", nvDom.style.color);
    });
    console.log("render");
    return (
      <div
        id="content"
        style={{ color: value === 1 ? "red" : "" }}
        onClick={() => setValue(c => c + 1)}
      >
        value: {value}
      </div>
    );
  };
```

当上述组件初始化后，在打印 render 后会打印一次 color effect，表明组件渲染之后，执行了传入的 effect。而在单击 ID 为 content 的元素后，将更新 value 状态，触发一次渲染，打印 render 之后会打印 color effect red。这一流程表明 React 的 DOM 已经更新完毕，并将控制权交给开发者的副作用函数，副作用函数成功地获取到了 DOM 更新后的值。事实上，上述流程与 React 的 componentDidMount、componentDidUpdate 生命周期类似，React 首次渲染和之后的每次渲染都会调用一遍传给 useEffect 的函数，这也是 useEffect 与传统类组件可以类比的地方。一般来说，useEffect 可类比为 componentDidMount、componentDidUpdate、componentWillUnmount 三者的集合，但要注意它们不完全等同，主要区别在于 componentDidMount 或 componentDidUpdate 中的代码是"同步"执行的。这里的"同步"指的是副作用的执行将阻碍浏览器自身的渲染，如有时候需要先根据 DOM 计算出某个元素的尺寸再重新渲染，这时候生命周期方法会在浏览器真正绘制前发生。

而 useEffect 中定义的副作用函数的执行不会阻碍浏览器更新视图，也就是说这些函数是异步执行的。所谓异步执行，指的是传入 useEffect 的回调函数是在浏览器的"绘制"阶段之后触发的，不"同步"阻碍浏览器的绘制。在通常情况下，这是比较合理的，因为大多数的副作用都没有必要阻碍浏览器的绘制。对于 useEffect，React 使用了一种特殊手段保证 effect 函数在"绘制"阶段后触发：

```
const channel = new MessageChannel();

channel.port1.onmessage = function() {
  // 此时绘制结束，触发 effect 函数
  console.log("after repaint");
```

```
};

requestAnimationFrame(function () {
  console.log("before repaint");
  channel.port2.postMessage(undefined);
});
```

requestAnimationFrame 与 postMessage 结合使用以达到这一类目的。

简而言之，useEffect 会在浏览器执行完 reflow/repaint 流程之后触发，effect 函数适合执行无 DOM 依赖、不阻碍主线程渲染的副作用，如数据网络请求、外部事件绑定等。

2．清除副作用

当副作用对外界产生某些影响时，在再次执行副作用前，应先清除之前的副作用，再重新更新副作用，这种情况可以在 effect 中返回一个函数，即 cleanup（清除）函数。

每个 effect 都可以返回一个清除函数。作为 useEffect 可选的清除机制，其可以将监听和取消监听的逻辑放在一个 effect 中。

那么，React 何时清除 effect？effect 的清除函数将会在组件重新渲染之后，并先于副作用函数执行。以一个例子来说明：

```
const App = () => {
  const [value, setValue] = useState(0);
  useEffect(function useEffectCallBack() {
    expensive();
    console.log("effect fire and value is", value);
    return function useEffectCleanup() {
      console.log("effect cleanup and value is ", value);
    };
  });
  return <div onClick={() => setValue(c => c + 1)}>value: {value}</div>;
};
```

每次单击 div 元素，都会打印：

```
// 第一次单击
effect cleanup and value is  0
effect fire and value is 1
// 第二次单击
effect cleanup and value is  1
```

```
effect fire and value is 2
// 第三次单击
effect cleanup and value is  2
effect fire and value is 3
// ……
```

如上例所示，React 会在执行当前 effect 之前对上一个 effect 进行清除。清除函数作用域中的变量值都为上一次渲染时的变量值，这与 Hooks 的 Caputure Value 特性有关，将在下面的内容中介绍。

除了每次更新会执行清除函数，React 还会在组件卸载的时候执行清除函数。

3. 减少不必要的 effect

如上面内容所说，在每次组件渲染后，都会运行 effect 中的清除函数及对应的副作用函数。若每次重新渲染都执行一遍这些函数，则显然不够经济，在某些情况下甚至会造成副作用的死循环。这时，可利用 useEffect 参数列表中的第二个参数解决。useEffect 参数列表中的第二个参数也称为依赖列表，其作用是告诉 React 只有当这个列表中的参数值发生改变时，才执行传入的副作用函数：

```
useEffect(() => {
  document.title = `You clicked ${count} times`;
}, [count]); // 只有当 count 的值发生变化时，才会重新执行 document.title 这一行
```

那么，React 是如何判断依赖列表中的值发生了变化的呢？事实上，React 对依赖列表中的每个值，将通过 Object.is 进行元素前后之间的比较，以确定是否有任何更改。如果在当前渲染过程中，依赖列表中的某一个元素与该元素在上一个渲染周期的不同，则将执行 effect 副作用。

注意，如果元素之一是对象或数组，那么由于 Object.is 将比较对象或数组的引用，因此可能会造成一些疑惑：

```
function App({config}) {
  React.useEffect(
    () => {},
    [config],
  )
  return <div>{/* UI */}</div>
}
// 每次渲染都传入 config 新对象
```

```
<App config={a:1}/>
```

如果 config 每次都由外部传入，那么尽管 config 对象的字段值都不变，但由于新传入的对象与之前 config 对象的引用不相等，因此 effect 副作用将被执行。要解决此种问题，可以依赖一些社区的解决方案，如 use-deep-compare-effect。

在通常情况下，若 useEffect 的第二个参数传入一个空数组[]（这并不属于特殊情况，它依然遵循依赖列表的工作方式），则 React 将认为其依赖元素为空，每次渲染比对，空数组与空数组都没有任何变化。React 认为 effect 不依赖于 props 或 state 中的任何值，所以 effect 副作用永远都不需要重复执行，可理解为 componentDidUpdate 永远不会执行。这相当于只在首次渲染的时候执行 effect，以及在销毁组件的时候执行 cleanup 函数。要注意，这仅是便于理解的类比，对于第二个参数传入一个空数组[]与这类生命周期的区别，可查看下面的注意事项。

4. 注意事项

1）Capture Value 特性

注意，React Hooks 有着 Capture Value 的特性，每一次渲染都有它自己的 props 和 state：

```
function Counter() {
const [count, setCount] = useState(0);
useEffect(() => {
const id = setInterval(() => {
console.log("count is", count);
setCount(count + 1);
}, 1000);
return () => clearInterval(id);
}, []);
return <h1>{count}</h1>;
}
```

在 useEffect 中，获得的永远是初始值 0，将永远打印"count is 0"；h1 中的值也将永远为 setCount(0+1)的值，即"1"。若希望 count 能依次增加，则可使用 useRef 保存 count，useRef 将在 3.2.4 节介绍。

2）async 函数

useEffect 不允许传入 async 函数，如：

```
useEffect(async () => {
// return 函数将不会被调用
```

```
}, []);
```

原因在于 async 函数返回了 promise，这与 useEffect 的 cleanup 函数容易混淆。在 async 函数中返回 cleanup 函数将不起作用，若要使用 async 函数，则可进行如下改写：

```
useEffect(() => {
  (async () => {
    // 一些逻辑
  })();
  // 可返回 cleanup 函数
}, []);
```

3）空数组依赖

注意，useEffect 传递空数组依赖容易产生一些问题，这些问题通常容易被忽视，如以下示例：

```
function ChildComponent({ count }) {
  useEffect(() => {
    console.log("componentDidMount", count);
    return () => {
      // 永远为 0，由 Capture Value 特性所导致
      alert("componentWillUnmount and count is " + count);
    };
  }, []);
  console.log("count", count);
  return <>count:{count}</>;
}
const App = () => {
  const [count, setCount] = useState(0);
  const [childShow, setChild] = useState(true);
  return (
    <div onClick={() => setCount(c => c + 1)}>
      <button onClick={() => setChild(false)}>销毁 Child 组件</button>
      {childShow && <ChildComponent count={count} />}
    </div>
  );
};
```

单击"销毁 Child 组件"按钮，浏览器将弹出"componentWillUnmount and count is 0"提示

框，无论 setCount 被调用多少次，都将如此，这是由 Capture Value 特性所导致的。而类组件的 componentWillUnmount 生命周期可从 this.props.count 中获取到最新的 count 值。

在使用 useEffect 时，注意其不完全与 componentDidUpdate、componentWillUnmount 等生命周期等同，应该以"副作用"或状态同步的方式去思考 useEffect。但这也不代表不建议使用空数组依赖，需要结合上下文场景决定。与其将 useEffect 视为一个功能来经历 3 个单独的生命周期，不如将其简单地视为一种在渲染后运行副作用的方式，可能会更有帮助。

useEffect 的设计意图是关注数据流的改变，然后决定 effect 该如何执行，与生命周期的思考模型需要区分开。

3.2.3　useLayoutEffect

React 还提供了与 useEffect 同等地位的 useLayoutEffect。useEffect 和 useLayoutEffect 在副作用中都可获得 DOM 变更后的属性：

```
const App = () => {
  const [value, setValue] = useState(0);
  useEffect(function useEffectCallBack() {
    const nvDom = document.getElementById("content");
    console.log("color effect", nvDom.style.color);
  });
  useLayoutEffect(function useLayoutEffectCallback() {
    const nvDom = document.getElementById("content");
    console.log("color layout effect", nvDom.style.color);
  });

  return (
    <div
      id="content"
      style={{ color: value === 1 ? "red" : "" }}
      onClick={() => setValue(c => c + 1)}
    >
      value: {value}
    </div>
  );
};
```

单击按钮后会打印"color layout effect red""color effect red"。可见 useEffect 与 useLayoutEffect 都可从 DOM 中获得其变更后的属性。

从表面上看，useEffect 与 useLayoutEffect 并无区别，但事实上厘清它们的区别需要从副作用的"同步""异步"入手。3.2.2 节曾介绍过 useEffect 的运行过程是异步进行的，即 useEffect 不阻碍浏览器的渲染；useLayoutEffect 与 useEffect 的区别是 useLayoutEffect 的运行过程是"同步"的，其阻碍浏览器的渲染。

简而言之，useEffect 发生在浏览器 reflow/repaint 操作之后，如果某些 effect 是从 DOM 中获得值的，如获取 clientHeight、clientWidth，并需要对 DOM 进行变更，则可以改用 useLayoutEffect，使得这些操作在 reflow/repaint 操作之前完成，这样有机会避免浏览器花费大量成本，多次进行 reflow/repaint 操作。以一个例子来说明：

```
const App = () => {
  const [value, setValue] = useState(0);
  useEffect(function useEffectCallBack() {
    console.log("effect");
  });
  // 在下一帧渲染前执行
  window.requestAnimationFrame(() => {
    console.log("requestAnimationFrame");
  });
  useLayoutEffect(function useLayoutEffectCallback() {
    console.log("layoutEffect");
  });
  console.log("render");
  return <div onClick={() => setValue(c => c + 1)}>value: {value}</div>;
};
```

分别在 useEffect、requestAnimationFrame、useLayoutEffect 和 render 过程中进行调试打印，以观察它们的时序。可以看到，当渲染 App 后将按如下顺序打印：render、layoutEffect、requestAnimationFrame、effect。由此可知，useLayoutEffect 的副作用都在"绘制"阶段前，useEffect 的副作用都在"绘制"阶段后。通过浏览器调试工具观察 task 的执行，如图 3-2 所示。

在图 3-2 中，①执行了 useLayoutEffectCallback，为 useLayoutEffect 的副作用；②为浏览器的 Paint 流程；在 Paint 流程后，③的执行函数为 useEffectCallBack，执行了 useEffect 的副作用。

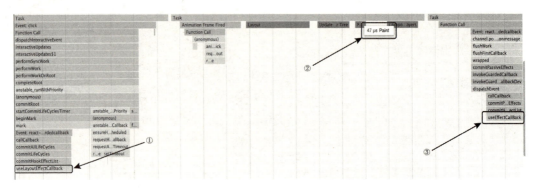

图 3-2　useLayoutEffectCallback 与 useEffectCallBack 的执行顺序

3.2.4　useRef

在使用 class 类组件时，通常需要声明属性，用以保存 DOM 节点。借助 useRef，同样可以在函数组件中保存 DOM 节点的引用：

```
import { useRef } from "React"
function App() {
    const inputRef = useRef(null);
    return <div>
        <input type="text" ref={inputRef}/>
        {/* 通过 inputRef.current 获取节点 */}
        <button onClick={() => inputRef.current.focus()}>focus</button>
    </div>
}
// useRef 的签名为：
interface MutableRefObject<T> {
    current: T;
}
function useRef<T>(initialValue: T): MutableRefObject<T>;
```

useRef 返回一个可变的 Ref 对象，其 current 属性被初始化为传递的参数（initialValue）。useRef 返回的可变对象就像一个"盒子"，这个"盒子"存在于组件的整个生命周期中，其 current 属性保存了一个可变的值。

useRef 不仅适用于 DOM 节点的引用，类似于类上的实例属性，useRef 还可用来存放一些与 UI 无关的信息。useRef 返回的可变对象，其 current 属性可以保存任何值，如对象、基本类型或函数等。所以，函数组件虽然没有类的实例，没有"this"，但是通过 useRef 依然可以解决数据的存储问题。如在 2.1 节，曾使用过 useRef：

```
function Example(props){
    const { history } = props;
    // 使用 useRef 保存注销函数
    const historyUnBlockCb = React.useRef<UnregisterCallback>(()=>{});
    React.useEffect(()=>{
        return ()=>{
            // 在销毁组件时调用，注销 history.block
            historyUnBlockCb.current()
        }
    },[]);
    function block() {
        // 解除之前的阻止
        historyUnBlockCb.current();
        // 重新设置弹框确认，更新注销函数，单击"确定"按钮，正常跳转；单击"取消"
        // 按钮，跳转不生效
        historyUnBlockCb.current = history.block('是否继续？')
    }
    return <>
        <button onClick={block}>阻止跳转</button>
        <button onClick={()=>{
            historyUnBlockCb.current();
        }}>解除阻止</button>
    </>
}
```

上例使用 useRef 返回了可变对象 historyUnBlockCb，通过 historyUnBlockCb.current 保存了 history.block 的返回值。

注意，更改 refObject.current 的值不会导致重新渲染。如果希望重新渲染组件，则可使用 useState，或者使用某种 forceUpdate 方法。

3.2.5 useMemo

作为 React 内置的 Hooks，useMemo 用于缓存某些函数的返回值。useMemo 使用了缓存，可避免每次渲染都重新执行相关函数。useMemo 接收一个函数及对应的依赖数组，当依赖数组中的一个依赖项发生变化时，将重新计算耗时函数。

```
function App() {
  const [count, setCount] = React.useState(0);
  const forceUpdate = useForceUpdate();
  const expensiveCalcCount = count => {
    console.log("expensive calc");
    let i = 0;
    while (i < 9999999) i++;
    return count;
  };
  // 使用 useMemo 记录高开销的操作
  const letterCount = React.useMemo(() => expensiveCalcCount(count), [count]);
  console.log("component render");
  return (
    <div style={{ padding: "15px" }}>
      <div>{letterCount}</div>
      <button onClick={() => setCount(c => c + 1)}>改变count</button>
      <button onClick={forceUpdate}>更新</button>
    </div>
  );
}
```

在上面的示例中,除了使用了 React.useState,还使用了一个自定义 Hook——useForceUpdate,其返回了 forceUpdate 函数,与类组件中的 forceUpdate 函数功能一致。关于自定义 Hook,将在 3.2.7 节介绍。

在初始渲染 App 时,React.useMemo 中的函数会被计算一次,对应的 count 值与函数返回的结果都会被 useMemo 记录下来。

若单击"改变 count"按钮,由于 count 改变,当 App 再次渲染时,React.useMemo 发现 count 有变化,将重新调用 expensiveCalcCount 并计算其返回值。因此,控制台会打印"expensive calc" "component render"。

而若单击"更新"按钮,则调用 forceUpdate 函数再次渲染。由于在再次渲染过程中 React.useMemo 发现 count 值没有改变,因此将返回上一次 React.useMemo 中函数计算得到的结果,渲染 App 控制台仅打印"component render"。

同时,React 也提供了 useCallback 用以缓存函数:

```
useCallback(fn, deps)
```

在实现上,useCallback 等价于 useMemo(() => fn, deps),因此这里不再赘述。

3.2.6 useContext

若希望在函数组件中使用 3.1 节中所述的 Context，除使用 Context.Consumer 消费外，还可使用 useContext：

```
const contextValue = useContext(Context);
```

useContext 接收一个 Context 对象（React.createContext 的返回值）并返回该 Context 的当前值。与 3.1 节中的 Consumer 类似，当前的 Context 值由上层组件中距离最近的 Context.Provider 提供。当更新上层组件中距离最近的 Context.Provider 时，使用 useContext 的函数组件会触发重新渲染，并获得最新传递给 Context.Provider 的 value 值。

调用了 useContext 的组件总会在 Context 值变化时重新渲染，这个特性将会经常使用到。

在函数组件中，使用 useContext 获取上下文内容，有效地解决了之前 Provider、Consumer 需要额外包装组件的问题，且由于其替代了 Context.Consumer 的 render props 写法，这将使得组件树更加简洁。

3.2.7 自定义 Hook

自定义 Hook 是一个函数，其名称约定以 use 开头，以便可以看出这是一个 Hooks 方法。如果某函数的名称以 use 开头，并且调用了其他 Hooks，就称其为一个自定义 Hook。自定义 Hook 就像普通函数一样，可以定义任意的入参与出参，唯一要注意的是自定义 Hook 需要遵循 Hooks 的基本准则，如不能在条件循环中使用、不能在普通函数中使用。

自定义 Hook 解决了之前 React 组件中的共享逻辑问题。通过自定义 Hook，可将如表单处理、动画、声明订阅等逻辑抽象到函数中。自定义 Hook 是重用逻辑的一种方式，不受内部调用情况的约束。事实上，每次调用 Hooks 都会有一个完全隔离的状态。因此，可以在一个组件中使用两次相同的自定义 Hook。下面是两个常用自定义 Hook 的示例：

```
// 获取 forceUpdate 函数的自定义 Hook
export default function useForceUpdate() {
  const [, dispatch] = useState(Object.create(null));
  const memoizedDispatch = useCallback(() => {
    // 引用变化
    dispatch(Object.create(null));
```

```
    }, [dispatch]);
    return memoizedDispatch;
}
```

获取某个变量上一次渲染的值：

```
// 获取上一次渲染的值
function usePrevious(value) {
  const ref = useRef();
  useEffect(() => {
    ref.current = value;
  }, [value]);
  return ref.current;
}
```

可基于基础的 React Hooks 定义许多自定义 Hook，如 useLocalStorage、useLocation、useHistory（将在第 5 章中进行介绍）等。将逻辑抽象到自定义 Hook 中后，代码将更具有可维护性。

3.3 Refs

3.3.1 createRef

前文曾介绍过 useRef 用以保存 DOM 节点，事实上也可以通过 createRef 创建 Ref 对象：

```
class MyComponent extends React.Component {
  constructor(props) {
    super(props);
    this.myRef = React.createRef();
  }
  render() {
    return <div ref={this.myRef} />;
  }
}
```

当 this.myRef 被传递给 div 元素时，可通过以下方式获取 div 原生节点：

```
const node = this.myRef.current;
```

Ref 不仅可以作用于 DOM 节点上，也可以作用于类组件上。在类组件上使用该属性时，Ref 对象的 current 属性将获得类组件的实例，因而也可以调用该组件实例的公共方法。

3.3.2 forwardRef

引用传递（Ref forwading）是一种通过组件向子组件自动传递引用 Ref 的技术。例如，某些 input 组件需要控制其 focus，本来是可以使用 Ref 来控制的，但是因为该 input 已被包裹在组件中，所以这时就需要使用 forwardRef 来透过组件获得该 input 的引用。

```
import React,{Component} from 'react';
import ReactDOM,{render} from 'react-dom';
const ChildOrigin = (props,ref)=>{
   return (
      <div ref={ref}>{props.txt}</div>
   )
};
const Child = React.forwardRef(ChildOrigin);
class Parent extends Component{
   constructor(){
      super();
      this.myChild = React.createRef();
   }
   componentDidMount(){
      console.log(this.myChild.current);// 获取的是Child组件中的div元素
   }
   render(){
      return <Child ref={this.myChild} txt="parent props txt"/>
   }
}
```

当对原 ChildOrigin 组件使用 forwardRef 获得了新的 Child 组件后，新 Child 组件的 Ref 将传递到 ChildOrigin 组件内部。在上面的示例中，可通过新 Child 组件的 Ref 值 this.myChild. current 获取到 ChildOrigin 组件内部 div 元素的引用。

3.4 Memo

为了提高 React 的运行性能，React v16.6.0 提供了一个高阶组件——React.memo。当 React.memo 包装一个函数组件时，React 会缓存输出的渲染结果，之后当遇到相同的渲染条件时，会跳过此次渲染。与 React 的 PureComponent 组件类似，React.memo 默认使用了浅比较的缓存策略，但 React.memo 对应的是函数组件，而 React.PureComponent 对应的是类组件。React.memo

的签名如下：

```
function memo<P extends object>(
    Component: SFC<P>,
    propsAreEqual?: (prevProps: Readonly<PropsWithChildren<P>>,
nextProps: Readonly<PropsWithChildren<P>>) => boolean
): NamedExoticComponent<P>;
```

React.memo 参数列表中的第一个参数接收一个函数组件，第二个参数表示可选的 props 比对函数。React.memo 包装函数组件后，会返回一个新的记忆化组件。以一个示例来说明，若有一个子组件 ChildComponent，没有通过 React.memo 记忆化：

```
function ChildComponent({ count }) {
  console.log("childComponent render", count);
  return <>count:{count}</>;
}
const App = () => {
  const [count] = useState(0);
  const [childShow, setChild] = useState(true);
  return (
    <div>
      <button onClick={() => setChild(c => !c)}>隐藏/展示内容</button>
      {childShow && <div>内容</div>}
      <ChildComponent count={count} />
    </div>
  );
};
```

当重复单击按钮时，由于触发了重新渲染，ChildComponent 将得到更新，将多次打印 "childComponent render"。若引入 React.memo(ChildComponent)缓存组件，则在渲染组件时，React 将进行检查。如果该组件渲染的 props 与先前渲染的 props 不同，则 React 将触发渲染；反之，如果 props 前后没有变化，则 React 不执行渲染，更不会执行虚拟 DOM 差异检查，其将使用上一次的渲染结果。

```
function ChildComponent({ count }) {
  console.log("childComponent render");
  return <>count:{count}</>;
}
const MemoChildComponent = React.memo(ChildComponent);
const App = () => {
```

```
  const [count] = useState(0);
  const [childShow, setChild] = useState(true);
  return (
    <div>
      <button onClick={() => setChild(c => !c)}>隐藏/展示内容</button>
      {childShow && <div>内容</div>}
      <MemoChildComponent count={count} />
    </div>
  );
};
```

当单击"隐藏/展示内容"按钮时，会导致重新渲染，但由于原组件通过 React.memo 包装过，使用了包装后的组件 MemoChildComponent，在多次渲染时 props 没有变化，因此这时不会多次打印"childComponent render"。

同时，React.memo 可以使用第二个参数 propsAreEqual 来自定义渲染与否的逻辑：

```
const MemoChildComponent = React.memo(ChildComponent, function propsAreEqual(
  prevProps,
  nextProps
) {
  return prevProps.count === nextProps.count;
});
```

propsAreEqual 接收上一次的 prevProps 与即将渲染的 nextProps，函数返回的 boolean 值表明前后 props 是否相等。若返回"true"，则认为前后 props 相等；反之，则认为不相等，React 将根据函数的返回值决定组件的渲染情况（与 shouldComponentUpdate 类似）。因此，可认为函数返回"true"，props 相等，不进行渲染；函数返回"false"则认为 props 有变化，React 会执行渲染。

注意，不能把 React.memo 放在组件渲染过程中。

```
const App = () => {
  // 每次都获得新的记忆化组件
  const MemoChildComponent = React.memo(ChildComponent);
  const [count] = useState(0);
  const [childShow, setChild] = useState(true);
  return (
    <div>
      <button onClick={() => setChild(c => !c)}>隐藏/展示内容</button>
```

```
        {childShow && <div>内容</div>}
        <MemoChildComponent count={count} />
    </div>
  );
};
```

这相当于每次渲染都开辟一块新的缓存，原缓存无法得到利用，React.memo 的记忆化将失效，开发者需要特别注意。

3.5 小结

本章介绍了 Context、Hooks、Refs、Memo 等 React 特性，在 React Router 源码及相关第三方库实现中，都涉及以上特性。掌握以上特性，对理解 React Router 及使用 React Router 进行实战都有非常大的帮助。

相比 props 和 state，React 的 Context 特性可以实现跨层级的组件通信。我们可以在很多框架设计中找到使用 Context 的例子，React Router 也是其一。学习使用 Context 对理解 React Router 十分重要。同时，本章介绍了 React Hooks，作为 React v16.8 的新特性，以及考虑到 React Router 今后的演进趋势，学习使用 React Hooks 进行函数式组件开发将对读者有极大的帮助。

参考文献

[1] https://zh-hans.reactjs.org/docs/context.html.

[2] https://en.wikipedia.org/wiki/Side_effect_(computer_science).

[3] https://zh-hans.reactjs.org/docs/hooks-reference.html#usecontext.

[4] https://github.com/facebook/react/blob/v16.8.6/packages/shared/shallowEqual.js.

[5] https://developer.mozilla.org/en-US/docs/Web/JavaScript/Reference/Global_Objects/Object/is#Description.

第 4 章
认识 React Router

依赖于浏览器 HTML 5 Session history 特性的普及，前端路由方案开始兴起，既有 Angular 的路由配置，也有与 Vue 搭配的 Vue Router。本章主要介绍 React Router 的起源、历史背景及技术发展脉络，提供学习 React Router 所需的基础知识，同时介绍 React Router 是什么、能发挥什么作用、为什么要使用 React Router 等。通过阅读本章内容，读者将对 React Router 有一个基本的了解。同时，本章借助一个工程应用的实现，为读者剖析 React Router 是如何融入 React 生态并发挥作用的。通过各模块循序渐进的引入，初学者能快速上手 React Router 的基本使用，并能理解 React Router 的三大基本要素，为后续内容的介绍做好铺垫。

4.1 React Router 是什么

在互联网演化进程中，早期网站中的内容多是静态内容，以文字、图片为主。随着互联网技术的发展和 HTML 5 技术的普及，现代的网页交互效果得到了显著提高。在前端路由方案还没有问世之前，页面的规划管理大多数是由服务端进行控制的，前端仅需要写好各个独立页面，并把页面交由服务端，由服务端进行页面的管理和维护。路由的方案多是使用超链接跳转的形式实现的。这种服务端提供页面的形式，在前端每次导航时，都会新开一个窗口或者刷新当前页面。

自 HTML 5 Session history 技术得到普及后，前端管理路由的方案开始兴起，前后端分离的方案也逐渐被广泛接受。路由管理逐渐从服务端过渡到前端，减少了服务端的关注点，在很多场景下服务端不再需要为前端提供页面访问。

React 是一个能将数据渲染为 HTML 视图的开源 JavaScript 库。自 2013 年 React 开源问世以来，其以声明式设计、组件化的开发、类 HTML 的 JSX 语法、高效的性能、代码逻辑简单等特点，越来越多地被人们关注和使用。同时，随着 React 的不断迭代更新，React 的生态也在蓬勃发展，并相继衍生出了许多优秀的开源 React 组件，基于 React 技术栈的路由管理工具 React Router 就是其一。

React Router 依托 React 强大的用户界面构建能力及浏览器 HTML 5 基础能力，以其强大的易用性，可以快速为基于 React 的项目引入路由管理能力，包括声明前端页面、管理页面间跳转等。自 2015 年 React Router v1.0.0 问世以来，各类项目均不同程度地使用 React Router。

除了 React Router，还有可与 Vue 搭配使用的 Vue Router、Angular 的路由管理系统、轻量版的@reach/router、与 UI 无关的路由状态管理工具 router 5 等都是优秀的路由管理工具，学习 React Router 也为掌握其他路由框架提供了理论及实践基础。

由于 React Router 基于 React，因此阅读本书需要一定的 React 基础，相关内容可参阅本书第 3 章。

4.2　React Router 版本的演进

React Router 经历过几次版本的演进，截至 2019 年年底，React Router 已发布到了第 5 版。在 React Router 版本迭代过程中，从 React Router v3.x 及之前版本的静态路由，到 React Router v4.x 及以后版本的动态路由，React Router 发生了很大的变化，对一些 API 接口进行了废弃，并有新的接口引入。本书以 React Router v5.x 为例进行介绍，同时兼容 React Router v4.x，还会在附录部分介绍 React Router v3.x 及之前版本的情况。

在 React Router v3.x 中，包括导航组件、路由匹配逻辑、导航 history 对象、命中逻辑等，都由 react-router 包提供；而到了 React Router v4.x 及以后的版本，原 react-router 包中的 DOM 部分拆分到了 react-router-dom 包中，如 Link、NavLink、BrowserRouter、HashRouter 等（今后也有合并到 react-router 包中的可能，如提供 react-router/dom）。若希望引入 Link 等组件，则需要从 react-router-dom 包中引入。在 React Router v5.x 中，react-router-dom 包依赖 react-router 包。react-router-dom 包与 react-router 包的关系为，react-router-dom 包是 react-router 包的超集，react-router-dom 包中具备了一切 react-router 包中的方法或组件。在一些情况下，React Router 的组件或方法既可以从 react-router-dom 包中引入，也可以从 react-router 包中引入。本书如无特别说明，均视作从 react-router 包中引入相关组件或方法。

4.3 静态路由与动态路由

在 React Router v3.x 及之前的版本中，路由为静态路由。所谓静态路由，即路由信息在开发阶段或者在程序运行前便已确立。这类提前设定好的静态路由，由于路由的结构信息完整，大多数可通过阅读源码或代码分析等手段分析出全部路由声明。在通常情况下，若要加载静态路由，则仅需把编写好的路由配置加载进内存即可。比如在 React Router v3.x 中，在 Router 组件的 componentWillMount 阶段，便收集了静态路由的信息并存入内存；之后，路由信息便不再更新，可以说 Route 在声明确立后便不能修改。同时，在现代前端体系的编译结果上，静态路由信息也可从编译后的 JavaScript 文件中获知。静态路由由于限制较大，缺少了灵活性，但是其具有便于代码静态分析、统一维护管理等特点。

在 React Router v4.x 和 React Router v5.x 中，React Router 舍弃了静态化的路由配置，改变为动态路由。相比静态路由，动态路由在运行时才能确立，且可以在运行时任意更新。由于此种动态特性，动态路由的路由信息很难通过静态代码分析等手段分析得出，无法从代码分析中分析出全部路由构造。动态路由抛弃了静态路由的规则限制，将更加灵活。对于 React Router v4.x 之后的动态路由组件，其更像能应对外部副作用的 React 组件，拥有组件的一切特性，能自由组合，灵活性高，自由度高，抛弃了配置的特性，真正成为可自由使用的 React 组件。

从 React Router 的迭代历程来看，从 React Router v3.x 到 React Router v4.x，变化最大的是路由的性质。在 React Router v3.x 及之前的版本中，静态路由的设计与 React 的理念有些脱节，即路由组件不像 React 组件。尽管路由组件 Route 在使用上类似于一个 React 组件，但是其却并不负责渲染，不像真正的 React 组件一样可灵活地与其他组件组合。路由组件 Route 不负责任何实质内容，仅作为配置，由 React Router 将配置加载到内存中。虽然这在一定程度上便于维护，但在编写 Route 时需要意识到路由 Route 仅提供一个配置，不具有组件应该具有的渲染元素的功能，且 React Router v3.x 所提供的生命周期方法在一定程度上重复了 React 组件的生命周期。React Router 意识到了这个问题，在 React Router v4.x 及以后的版本中，一是将静态路由改为了动态路由，二是使得 React Router 更加贴合 React 的组件化设计，还原了组件应有的功能，使得路由组件化，如路由组件的渲染、组合拼装等，而不是配置化的声明且不能随意组合，这使得 React Router 与 React 更加靠齐。比如在 React 页面编写过程中，使用 React Router v4.x 可先忽略应用的路由结构，按页面结构划分设计好相关的页面组件，利用动态路由可组件化的特性，在设计好页面组件后，可在任何希望路由化的组件外层使用 React Router，使得组件的渲染与路由路径相关联。

需要注意的是，静态路由与动态路由都是一种实现方式。在类似的路由框架中，如 Angular、Vue Router，使用的都是静态路由。静态路由有着便于维护等特点，而动态路由灵活性高，我们应在具体场景中具体分析该使用何种路由类型。本书也提供 React Router 动态路由转换为静态路由的配置化方法。

同时，对于一些旧有项目来说，如使用 React Router v3.x，若希望迁移到 React Router v4.x 及以上版本，从静态路由升级到动态路由，则可以查看本书的附录 A 部分。

4.4 使用 React Router 实现一个工程应用

为了便于读者快速掌握 React Router 的全貌，以及对 React Router 有进一步的认识，本节将为读者呈现一个较为简单的应用。如图 4-1 所示，应用有 3 个页面，分别为商家列表页、商家详情页和美食详情页。其中，商家列表页列出了一系列的商家信息，单击相关商家会跳转到对应商家的商家详情页。在商家详情页中，有这个商家所提供的所有美食推荐菜，单击相关的美食推荐菜会跳转到美食详情页，该页面会介绍美食的相关情况及众多的推荐菜评价。

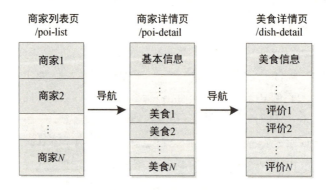

图 4-1　应用页面及导航

通过页面路由的声明及页面间的跳转导航，在本例中将会看到 React 组件是如何与路由结合的。从目录结构开始，声明如下的目录结构：

```
|    index.tsx
|    styles.css
|
├─pages
|        dishDetail.tsx
|        poiDetail.tsx
|        poiList.tsx
|
└─services
    |   index.ts
    |
    └─mockData
            poiList.ts
```

index.tsx 作为应用的入口，需要负责整个应用的渲染；style.css 提供应用的样式信息。pages 提供了应用页面的 React 组件，分别有 PoiList 商家列表页组件、PoiDetail 商家详情页组件和 DishDetail 美食详情页组件。services 目录用于提供页面所需的数据，这里使用了构造的数据，真实情况一般需要通过接口获取。

构造基本的 React 应用入口 index.tsx，引入 react 和 react-dom，并渲染一个 App 组件。

```
import * as React from "react";
import { render } from "react-dom";
import "./styles.css";

function App() {
  return (
    <div className="App">
      ……
    </div>
  );
}
const rootElement = document.getElementById("root");
render(<App />, rootElement);
```

为了能使路由与 React 应用相结合，需要安装并载入 react-router 包：

```
// 载入 react-router 包
npm install react-router
```

在载入 react-router 包之后，需要从 react-router 包中引入路由器 Router，以及从 history 库中

引入相关 history 的创建方法：

```
import { Router } from "react-router";
// createBrowserHistory 可查看 2.2.1 节
import { createBrowserHistory } from "history";
```

由于 react-router 包的依赖列表中包含了 history，因此 history 不需要再单独安装和载入。

通过导入 Router，引入路由能力的提供者，并且引入浏览器路径管理能力的提供者 history，为后续做准备。

在引入 createBrowserHistory 创建方法后，可创建对应的 history。这一步的目的是创建浏览器路径管理能力的提供者 history：

```
const history = createBrowserHistory();
```

如果从 react-router-dom 包中引入了 BrowserRouter，则这一步可以省略。

在引入各实例变量后，在 App 中实例化 Router：

```
function App() {
  return (
    <div className="App">
      <Router history={history}>
        ……
      </Router>
    </div>
  );
}
```

以上示例在顶层渲染了 Router 组件，且为了获得浏览器路径管理的能力，Router 的 history 属性传入了之前创建好的 history 对象。这一步通过渲染 Router 组件，使得应用拥有了路由管理能力。

为了实现 3 个不同页面间的跳转，每个页面都应有对应的 URL。根据图 4-1 的设计，商家列表页的 URL 为"/poi-list"，当用户打开相关域名下的/poi-list 路径时，即可看到对应的页面内容。商家详情页的 URL 为"/poi-detail"，这样单击某个具体商家后，将会导航到该 URL 下。美食详情页的 URL 为"/dish-detail"，当单击某个具体的美食时，会跳转到这个 URL 路径。

在设计好 3 个页面的 URL 路径后，为了将 pages 目录下的 React 组件与 URL 路径关联起来，需要引入 Route：

```
import { Route } from "react-router";
```

Route 的能力是将 URL 路径与 React 组件相关联，先从 pages 目录下引入相关的组件 PoiList、PoiDetail 和 DishDetail，并将 Route 结合路径 path 与组件 component 写在 Router 的任意子组件位置：

```
import * as React from "react";
import { render } from "react-dom";
import { Router, Route } from "react-router";
import { BrowserRouter } from "react-router-dom";
import { createBrowserHistory } from "history";
import PoiList from "./pages/poiList";
import PoiDetail from './pages/poiDetail';
import DishDetail from './pages/dishDetail';
const history = createBrowserHistory();
function App() {
  return (
    <div className="App">
      <Router history={history}>
        <Route path="/poi-list" component={PoiList} />
        <Route path="/poi-detail" component={PoiDetail} />
        <Route path="/dish-detail" component={DishDetail} />
      </Router>
    </div>
  );
}
const rootElement = document.getElementById("root");
render(<App />, rootElement);
```

Router 的子组件包含 3 个 Route 组件。其中，Route 的 path 属性为图 4-1 设计的期望展示的路径，component 属性表示该路径下希望渲染的 React 组件，即从 pages 目录下引入的 React 页面组件。通过渲染 Route，即路径/poi-list 展示 PoiList 商家列表页组件，/poi-detail 展示 PoiDetail 商家详情页组件，/dish-detail 展示 DishDetail 美食详情页组件。

通过 Route 的引入，当页面路径为/poi-list 时，即会渲染 PoiList 商家列表页组件：

```
// PoiList 组件
// 引入 RouteComponentProps 类型
import { RouteComponentProps } from "react-router";
import * as React from "react";
// 获取数据的方法
import { getPoiList } from "../services/index";
export default function PoiList(props: RouteComponentProps) {
  const [poiList, setPoiList] = React.useState([]);
```

```
  React.useEffect(() => {
    getPoiList().then(poiList => {
      // 获取商家列表数据并更新状态
      setPoiList(poiList);
    });
  }, []);
  return (
    <>
      <div>推荐商家</div>
      <div className="pois">
        {/* 展示商家列表 */}
        {poiList.map((poi, index) => {
          return (
            <div key={index}>
              <img src={poi.newImgUrl} />
              <div>{poi.shopName}</div>
            </div>
          );
        })}
      </div>
    </>
  );
}
```

使用 React v16.8 的 Hooks 写法实现了 PoiList 商家列表页组件。使用 useState 保存商家列表，初始商家列表为空数组，返回的 poiList、setPoiList 类似于 this.state 与 this.setState。当 useEffect 的第二个参数为空数组时，可认为其中的函数将在 componentDidMount 阶段执行。组件在挂载成功后调用 getPoiList 异步方法获得商家列表，随后调用 setPoiList 更新商家信息。

通过在浏览器中输入/poi-list 地址，可获得如图 4-2 所示的商家列表。

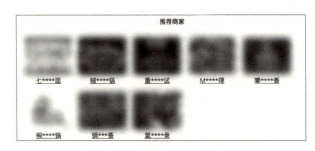

图 4-2　在/poi-list 路径下渲染商家列表

在完成商家列表渲染之后，需要为每个商家引入导航，使得可在单击每条商家信息后跳转到对应商家的详情页查看该商家所推荐的美食。这时可从 react-router-dom 包中引入 Link 组件，首先安装 react-router-dom 包：

```
npm install react-router-dom
```

安装成功后，可在文件 poiList.tsx 中引入：

```
import { Link } from "react-router-dom";
```

并将组件返回的部分更新为：

```
  return (
    <>
      <div>推荐商家</div>
      <div className="pois">
        {poiList.map((poi, index) => {
          // 引入 Link 组件用于导航
          return (
            <Link to={`/poi-detail?id=${poi.id}`} key={index}>
              <img src={poi.newImgUrl} />
              <div>{poi.shopName}</div>
            </Link>
          );
        })}
      </div>
    </>
  );
```

在上例中组件渲染每一个商家时，使用引入的 Link 组件作为外层组件。Link 的作用为导航，表明单击商家信息的任意一处都会导航到 Link 的 to 属性所指明的位置。由于在 Rotue 处曾声明过/poi-detail 路径，因此这里导航的路径为/poi-detail?id=poi.id。"?"后面的字符串为浏览器路径的 search 部分，表明需要携带的额外信息。这里将商家的 id 号作为额外信息带入/poi-detail 页面。/poi-detail 路径对应的文件为 poiDetail.tsx，当导航到/poi-detail 路径后，对应组件得到渲染：

```
// poiDetail.tsx 文件的内容
import { RouteComponentProps } from "react-router";
import * as React from "react";
import { getPoiDetailService } from "../services/index";
export default function PoiDetail(props: RouteComponentProps) {
  const [poiInfo, setPoiInfo] = React.useState(null);
```

```
React.useEffect(() => {
  const query = new URLSearchParams(props.location.search);
  getPoiDetailService(query.get("id")).then(poiInfo => {
    setPoiInfo(poiInfo);
  });
}, []);
return (
  poiInfo && (
    <>
      <div>商家详情</div>
      <div>地址：{poiInfo.address}</div>
      <div>电话：{poiInfo.phone}</div>
      <div>人均：{poiInfo.average}元</div>
      <div className="dishes">
        {poiInfo.dishes.map((dish, index) => {
          return (
            <div key={index}>
              <img className="dish-pic" src={dish.frontImgUrl} />
              <div>{dish.name}</div>
              <div>{dish.price}元</div>
            </div>
          );
        })}
      </div>
    </>
  )
);
}
```

  为了获得上一个跳转前页面/poi-list带入的商家id信息，这里使用了props的location.search。通过props.location.search可获得浏览器路径问号后面的search部分（在BrowserHistory情况下），并将search传入URLSearchParams进行实例化，即可通过实例化对象的get方法获得传入的商家id信息。利用得到的商家id信息，可调用对应的service方法，获取到具体商家的地址、联系电话、人均价格、美食推荐菜等信息。这里同样使用了React Hooks与人工构造的数据，如图4-3所示。

第 4 章 认识 React Router

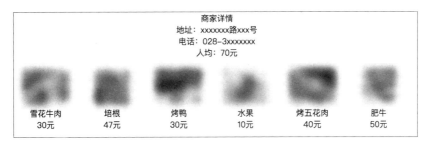

图 4-3 /poi-detail 商家详情页面

同理,为了进入商家具体菜品的详情页,也可使用上例中的 Link 组件导航,但若此时不希望使用 Link 组件导航,而是希望编程化导航并在导航前进行一些逻辑判断的操作,则可使用 history,将推荐菜列表改写为:

```
return (
  poiInfo && (
    <>
      <div>商家详情</div>
      <div>地址: {poiInfo.address}</div>
      <div>电话: {poiInfo.phone}</div>
      <div>人均: {poiInfo.average}元</div>
      <div className="dishes">
        {poiInfo.dishes.map((dish, index) => {
          return (
            <div
              onClick={() => {
                // 使用 history 导航
                props.history.push(`/dish-detail?dishId=${dish.id}`);
              }}
              key={index}
            >
              <img className="dish-pic" src={dish.frontImgUrl} />
              <div>{dish.name}</div>
              <div>{dish.price}元</div>
            </div>
          );
        })}
      </div>
    </>
```

```
    )
  );
```

上例监听 dishes 中每个渲染出的 dish 菜品的 onClick 函数，在函数中使用 props.history 对象（该对象即第 2 章所介绍的 browserHistory 对象）。类似 Link 的 to 属性，使用 props.history.push 方法进行导航跳转，方法的参数为要导航到的具体路径。使用 props.history.push 方法导航到 /dish-detail，/dish-detail 对应的文件为 dishDetail.tsx：

```
// dishDetail.tsx 文件的内容
import { RouteComponentProps } from "react-router";
import * as React from "react";
import { getDishDetailService } from "../services/index";
export default function DishDetail(props: RouteComponentProps) {
  const [dish, setDish] = React.useState(null);
  React.useEffect(() => {
    const query = new URLSearchParams(props.location.search);
    // 同样从 props.location.search 中获取到菜品 id 信息
    getDishDetailService(query.get("dishId")).then(dish => {
      setDish(dish);
    });
  }, []);
  return (
    dish && (
      <>
        <div className="dish">
          <img className="dish-img" src={dish.img} />
          <div>菜品名：{dish.name}</div>
          <div>推荐数：{dish.recommends}</div>
          <div>评分：{dish.rate}</div>
          <div>评论：{dish.comments.join(" ")}</div>
        </div>
      </>
    )
  );
}
```

/dish-detail 路径将渲染上例中的 React 组件，同样从 props.location.search 中获取到上一个 /poi-detail 页面所传入的菜品 id 信息，并通过调用相关接口方法请求到如菜品名等数据并渲染。/dish-detail 美食详情页面如图 4-4 所示。

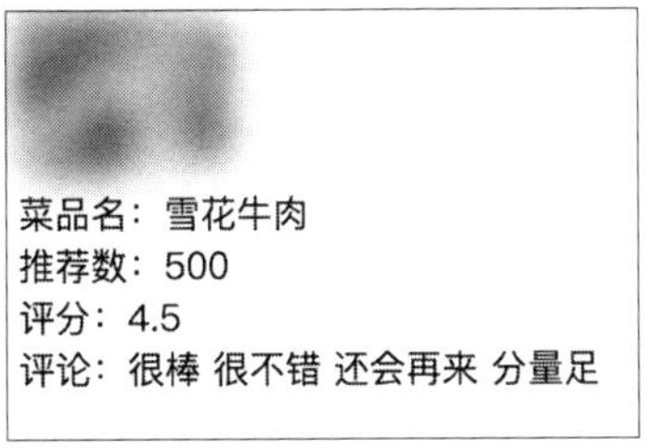

图 4-4 /dish-detail 美食详情页面

4.5 小结

从 4.4 节的例子中可以学习到，在 React 应用中使用的 React Router 包含三部分：Router 路由管理者组件、Route 路由端口组件及 Link 或 history 导航组件和方法。

对于 Router，其在 React 应用中作为路由信息的分发器，在任何一个 React Router 应用中都应将其作为顶层组件使用。4.4 节中使用的 BrowserRouter，其特点是与浏览器的路径系统兼容，在多数场景下建议使用此 Router；不过需要注意的是，BrowserRouter 在一般情况下需要服务端的支持才能运行，这里的服务端可以为 Nginx 或 Node 等。BrowserRouter 将在 5.3 节进行介绍。对于 Web 浏览器，除了 4.4 节中使用的 BrowserRouter，还可使用 HashRouter。HashRouter 使用浏览器的 hash 部分作为应用的路径，在 SPA 页面中兼容性较好，不过由于 React Router v4.x 的 history 库中 hashHistory 不支持 pushState 等操作，因而 HashRouter 不具备 history.state 状态的存储。但由于 HashRouter 使用了浏览器地址的 hash 部分，其运行并不需要服务端的支持，不需要额外的服务端配置，这也是其优点。HashRouter 将在 5.4 节进行介绍。除了在浏览器中运行 React Router，React Router 还提供了 StaticRouter、MemoryRouter 和 NativeRouter。StaticRouter 意为静态路由，也称为无状态路由，通常在服务端 Node.js 应用程序中使用，通过维护一个上下文对象保存导航信息，这将在 5.6 节介绍。MemoryRouter 和 NativeRouter 作为内存路由，所有路径信息都存储于内存中，它们与 BrowserRouter 的接口基本一致。MemoryRouter 和 NativeRouter 在测试场景或 React Native 场景中使用较多。

Route 组件的职责为接收路径信息并执行渲染。Route 也称为路由端口，用于接收 Router 的命令，每个 Route 都有自己的职责。我们将在第 6 章详细介绍 Route，并且在后面的章节中也会看到其他的用法。

导航作为路由变化的发起者，职责为发出路由变化的命令。在各类 Router 应用中，会根据不同的 Router（如 BrowserRouter 和 HashRouter）发出不同的命令，对应的 Router 也能响应这些命令。在 4.4 节中使用了 Link 组件与 history 方法，Link 组件会渲染 a 标签作为导航元素，history 方法曾在第 2 章中有过介绍。在第 7 章中还会介绍带激活态的 NavLink 组件，以及 React Native 的 DeepLinking 等导航组件，同时在第 8 章中会介绍重定向导航组件 Redirect。

第 5 章 Router

本章将全面介绍路由系统的第一个基本要素：Router。通过引入 Router 源码，本章将介绍 Router 的运行原理。学完本章，会为读者理解整个路由系统打好重要基础。同时，本章将对现有不同种类的 Router 进行归纳总结，以便读者能根据具体场景合理运用。

5.1 Router 是什么

Router 是 React Router 的导航命令传递者，若没有引入 Router，那么任何跳转都不会生效。Router 是一个与运行环境无关的 React 组件，其根据接收外部 history 对象的不同能提供不同的功能：如果接收 browserHistory，则得到 BrowserRouter，称为浏览器路由；如果接收 hashHistory，则得到 HashRouter，称为哈希路由；如果接收 memoryHistory，则得到 MemoryRouter，称为内存路由，如图 5-1 所示。

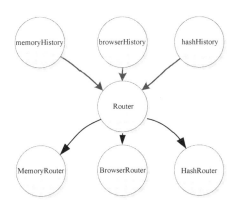

图 5-1　Router 可接收不同的 history 对象

5.2 Router 源码解析

Router 作为 React Router 顶层需要引入的组件,为应用提供了组件化的路由响应能力。学习 Router,将会知晓 React 应用是如何得知路由变化并做出响应的。Router 源码拆分后可以分为三部分:history 监听、提供初始 Context 及提前监听。

5.2.1 history 监听

第 2 章介绍的各 history 对象提供了 listen 函数,Router 使用了这一特性。由于 Router 并不需要关心 history 对象的类型,因此 Router 仅需要调用 history 的 listen 函数,感知路由的变化由 history 负责,Router 仅负责组件的逻辑:

```js
this.unlisten = props.history.listen(location => {
  this.setState({ location });
});
```

5.2.2 提供初始 Context

在 3.1 节中曾介绍过 React 的 Context 特性,在 React Router 中就大量运用了此特性提供路由响应支持。通过一级级提供的 Context,使得路由状态得以传递,顶层的 Context 由 Router 提供:

```js
// Router 源码
import React from "react";
import RouterContext from "./RouterContext";
class Router extends React.Component {
  static computeRootMatch(pathname) {
    return { path: "/", url: "/", params: {}, isExact: pathname === "/" };
  }
  constructor(props) {
    super(props);
    this.state = {
      // 初始 location 从 history 中获得
      location: props.history.location
    };
```

```
    this._isMounted = false;
    this._pendingLocation = null;
    if (!props.staticContext) {
      this.unlisten = props.history.listen(location => {
        if (this._isMounted) {
            // 仅在挂载成功后进行设置，防止出现未挂载成功而 props.history.location
            // 已经变化的情况
          this.setState({ location });
        } else {
            // 先将最近的 location 变化挂载
          this._pendingLocation = location;
        }
      });
    }
  }
  componentDidMount() {
      // 用于记录组件是否挂载成功
    this._isMounted = true;
    if (this._pendingLocation) {
        // 将挂载的 location 设置到 state.location 中
      this.setState({ location: this._pendingLocation });
    }
  }
  componentWillUnmount() {
      // unmount 在销毁时应取消监听
    if (this.unlisten) this.unlisten();
  }
  render() {
    return (
      <RouterContext.Provider
        children={this.props.children || null}
        value={{
          history: this.props.history,
            // state.location 发生了变化，触发消费者重新渲染
          location: this.state.location,
          match: Router.computeRootMatch(this.state.location.pathname),
          staticContext: this.props.staticContext
        }}
```

```
      />
    );
  }
}
export default Router;
```

源码使用了 RouterContext 的上下文，使用 RouterContext.Provider 提供跨组件的值传递。RouterContext 如下：

```
// RouterContext
// React.createContext 的兼容性方案
import createContext from "mini-create-react-context";
const createNamedContext = name => {
  const context = createContext();
  context.displayName = name;
  return context;
};
const context = /*#__PURE__*/ createNamedContext("Router");
export default context;
```

在 React Router 中，RouterContext 由第三方库 mini-create-react-context 提供。mini-create-react-context 作为 React.createContext 的兼容性方案，其暴露方法可用于创建与 React Router v16.3 中 React.createContext 相同接口的上下文，作为 React.createContext 的 polyfill，支持 React v15.x。对于 React v16.8 及之后的版本，mini-create-react-context 内部将使用 React 原生的 React.createContext 接口。

在 Router 使用了 props.history.listen 监听后，在回调函数中，setState 更新 location 信息并通过 RouterContext.Provider 通知各消费者。注意，Router 不能传入 location，所有的 location 更新都由回调函数触发。

在未发生路由跳转时，初始的 location 为 props.history.location。使用传入 history 对象所提供的初始值，如 browserHistory 的初始 location 为 window.location。

注意，初始 Router 的 match 由 computeRootMatch 提供：

```
static computeRootMatch(pathname) {
  return { path: "/", url: "/", params: {}, isExact: pathname === "/" }
}
```

所以，顶层 Router 所提供的 context.match 必定不为 null，这个特性会影响无路径 Route 的渲染，这将在第 6 章进行介绍。

5.2.3 提前监听

源码中有_isMounted 与_pendingLocation 内部维护的标识和对象，在 Router 未挂载时（未进入_componentDidMount 生命周期），如果浏览器历史发生变化，则将变化 location 记录到 pendingLocation 中，且 history.listen 的监听在 Router 组件的构造函数中，而非在 componentDidMount 生命周期中。这是为了兼容初始化场景下的 Redirect 子组件，由于当 Redirect 初始化挂载后，即会进行 history.push 操作。如果在 componentDidMount 生命周期中监听 history 的变化，则由于 Router 一般位于组件树的顶层，componentDidMount 将在最后得到执行，出现导航先于监听的错误情况。虽然 Router 还未挂载，但可监听到子组件的任意导航信息，并可缓存下来以便在 Router 挂载成功后使用。

5.3　BrowserRouter

BrowserRouter 使用 browserHistory 进行地址管理，如 4.4 节所述。BrowserRouter 可以从 react-router 包中引入 Router 并传入 browserHistory 实现，或者从 react-router-dom 包中引入 BrowserRouter：

```
// react-router-dom 包中的 BrowserRouter 源码
import React from "react";
import { Router } from "react-router";
import { createBrowserHistory as createHistory } from "history";
class BrowserRouter extends React.Component {
  history = createHistory(this.props);
  render() {
    return <Router history={this.history} children={this.props.children} />;
  }
}
export default BrowserRouter;
```

由于 BrowserRouter 使用了 browserHistory，因此其拥有 browserHistory 的一切特性，如 pushState/replaceState 状态存储、history 地址与浏览器地址栏地址兼容等。

对于使用 BrowserRouter 的应用，需要服务端提供一定的支持，不管是 Nginx 代理、Node.js

还是其他服务端程序。原因在于用户强制刷新场景。设想当前的路径为根路径，如果仅给用户提供单纯的 CDN 静态文件，那么考虑当使用 BrowserRouter 后如导航到/foo/baz 路径，这时用户强制刷新了页面，如果仅有单纯 CDN 静态文件的支持，由于找不到/foo/baz 路径下的资源，页面就会返回 "404 无法找到" 或者视具体情况产生其他错误。对于使用 BrowserRouter 的应用，在大多数场景下可使用服务端渲染用户主动刷新的部分视图，当服务端渲染工作完成后，之后的路由可由 BrowserRouter 独立完成（对于使用 React 的服务端渲染，其路由支持可查看 5.6 节）。也可使用 Nginx 等代理，将相关的路径（如以/web 开头的路径）都转发到静态文件上，当静态文件执行后，会读取当前的浏览器路径并正确渲染对应的组件。

5.4 HashRouter

HashRouter 用来为 React 应用提供 hash 路由支持，使用 HashRouter 的 React 应用不需要服务端的支持。HashRouter 使用的 history 为 hashHistory，路由路径为浏览器地址栏地址 "#" 后面的部分，hashHistory 的相关介绍可查阅 2.3 节。

与 BrowserRouter 类似，使用 HashRouter 也有两种方式，一种方式是从 react-router-dom 包中直接引入 HashRouter：

```
import {HashRouter} from "react-router-dom";
<HashRouter>
  <App />
</HashRouter>
```

另一种方式是从 react-router 包中引入 Router 后，再将 hashHistory 传入，与第一种方式是等价的：

```
import { Router} from "react-router";
import { createHashHistory } from "history";
const hashHistory = createHashHistory();
<Router history={hashHistory}>
  <App />
</Router>
```

由于 HashRouter 不支持 state 持久化存储，其目的是支持在旧式浏览器上运行路由，所以建议读者在现代浏览器中使用 BrowserRouter，以获得全面的浏览器特性支持。

5.5 NativeRouter

对于 NativeRouter，其使用场景为 React Native，其使用的 history 为 memoryHistory。NativeRouter 的使用方式如下：

```
import NativeRouter from "react-router-native";
```

对于 NativeRouter，可传入 memoryHistory 中的配置，如：

```
// 初始路径为"/"
<NativeRouter initialEntries={["/"]} />;
```

NativeRouter 内部使用的是 MemoryRouter（MemoryRouter 为使用了 memoryHistory 的 Router），并使用 react-native 的 Alert 组件作为默认的 getUserConfirmation 函数传入 MemoryRouter 中：

```
import React from "react";
import { Alert } from "react-native";
import { MemoryRouter } from "react-router";
function NativeRouter(props) {
  // 透传配置 props 到 MemoryRouter 中
  return <MemoryRouter {...props} />;
}
NativeRouter.defaultProps = {
  // 默认使用 react-native 的 Alert 组件，用于在 history.block 场景中提示用户
  getUserConfirmation: (message, callback) => {
    Alert.alert("Confirm", message, [
      { text: "Cancel", onPress: () => callback(false) },
      { text: "OK", onPress: () => callback(true) }
    ]);
  }
};
export default NativeRouter;
```

React Native 是使用 JavaScript 编写原生移动应用的框架，它在设计原理上和 React 一致，通过声明式的组件机制来搭建丰富多彩的用户界面。React Native 所使用的基础 UI 组件和原生应用完全一致，使用 React Native 框架开发的最终产品是一个真正的移动应用。

5.6 StaticRouter

StaticRouter 一般称为静态路由。StaticRouter 与其他类型 Router 的最大区别在于其不改变路径地址，且不记录历史栈，为无状态路由，大多在服务端渲染场景中使用。虽然 MemoryRouter 也不改变外部环境的地址，但是其使用内存模拟了历史栈，浏览的记录在内存中都真实存在，是有状态的。而 BrowserRouter 与 HashRouter 也间接地使用了浏览器自身的能力保留了历史栈，也是有状态的。在 5.1 节中曾介绍过，Router 的地址管理能力都由 history 提供，对于 StaticRouter 也是如此。但是其 history 并不由 history 第三方库提供，而是直接内化在其源码实现中。对于 StaticRouter，可从 react-router 包中引入，无须传入 history：

```
import { StaticRouter } from "react-router";
// 接收 location 与 context
<StaticRouter location={req.url} context={staticContext}>
......
</StaticRouter>
```

其接收的 props 为：

```
export interface StaticRouterProps {
  // 基准地址
  basename?: string;
  // 初始化地址
  location?: string | object;
  // 服务端的上下文对象
  context?: {
    url?: string;
    action?: 'PUSH' | 'REPLACE';
    location?: object;
    statusCode?: number;
  };
}
```

对于 basename，其子组件的所有导航项都会带上 basename 作为路径前缀：

```
<StaticRouter basename="/foo">
  <Redirect to="/baz"/> // 重定向到/foo/baz
</StaticRouter>
```

其接收的 props 相比其他类型的 Router，还能接收 location。locaiton 可为字符串或 location 对象，这一点是其他 Router 不允许的。同时还能接收一个 context 可变对象，这个可变对象用于保存路径等信息，一般初始时传入一个空对象，这个空对象的各属性值在后续业务流程的操作中会根据情况发生变化。StaticRouter 的源码较为简单，如下所示：

```
function noop() {}
class StaticRouter extends React.Component {
  navigateTo(location, action) {
    const { basename = "", context = {} } = this.props;
    // 更新 action
    context.action = action;
    // 更新 location，即 staticContext 的 location
    context.location = addBasename(basename, createLocation(location));
    // 更新 URL，即 staticContext 的 URL
    context.url = createURL(context.location);
  }
  handlePush = location => this.navigateTo(location, "PUSH");
  handleReplace = location => this.navigateTo(location, "REPLACE");
  render() {
    const { basename = "", context = {}, location = "/", ...rest } = this.props;
    const history = {
      createHref: path => addLeadingSlash(basename + createURL(path)),
      // 初始为 POP
      action: "POP",
      // 初始的 location 依赖外部 props 的传入
      location: stripBasename(basename, createLocation(location)),
      // 仅 push 或 replace 会改变 context 的 action、location、URL，当然也可以从
      // props.staticContext 中获取 context 进行改变
      // 注意这些操作都在服务端进行
      push: this.handlePush,
      replace: this.handleReplace,
      // 以下方法全部不做处理
      go: noop,
      goBack: noop,
      goForward: noop,
      // 对于 listen，Router 也做了判断，当没有传入 staticContext 对象时，才调用
      //listen 函数监听，否则不会调用
```

```
        listen: noop,
        // block 无效
        block: noop
      };
      // 将 staticContext 对象传入 Router，这时 Router 的行为将静态化，不响应 URL 变化，
      // 仅对初始加载做出渲染反应
      return <Router {...rest} history={history} staticContext={context} />;
    }
}
export default StaticRouter;
```

如源码所示，StaticRouter 不提供 listen、go、block 等方法，原因在于 StaticRouter 为无导航状态的。

类似于 MemoryRouter 的 initialEntries 配置，StaticRouter 也能配置初始的 location，只需将 location 传入 StaticRouter 中即可。传递 location 的意义在于初始化内部的 history.location，之后 history 的 location 将永不变化。这符合服务端渲染的无状态逻辑，如果有变更，则应更新到 context 上下文中。在 React 应用服务端使用 StaticRouter，即为产生并修改 context 上下文，并供后续业务流程消费使用。

由于 StaticRouter 在服务端使用，各组件调用 props.history.push 或 props.history.replace 方法也在服务端进行，因此这些操作仅会改变 context 对象中 action、location、URL 的值，对外部浏览器不会有影响。

通常有两种更新 context 上下文的方式：

- 在服务端运行时，通过组件调用 props.history.push 或 props.history.replace 方法改变 context 上下文的 action、location 和 URL。
- 在服务端运行时，从 Route 注入的 props 中获取上下文 context 对象 props.staticContext，可直接在该对象上操作，如设置某个属性的值。

下面以一个实例应用说明 StaticRouter 服务端渲染的用法，这里忽略 webpack 的配置，仅关注 StaticRouter 在服务端的使用。应用有 3 个文件：App.jsx、client.js 和 server.js。App.jsx 作为应用的真实渲染内容，由服务端与客户端共同使用。client.js 供客户端渲染使用，一般会打包成 JavaScript 文件。server.js 在服务端运行，并执行服务端渲染的逻辑。

```
// App.jsx 文件，由服务端与客户端共用，组件的 render 函数都将在服务端与客户端得到执行
import React, { Component } from "react";
import {Redirect,Route} from "react-router";
// 用于记录状态码的路由
```

```
function RouteWithNotFound(props) {
  return (
    <Route
      render={routeProps} => {
        // 如果在浏览器中没有这个对象,则需要进行判断并做出处理
        if (routeProps.staticContext) {
          // 当渲染这个组件时,表明未找到相关的渲染项,设置 404 状态码到上下文中
          // 在服务端可使用此状态码
          routeProps.staticContext.statusCode = 404;
          // 在服务端不渲染组件,设置完状态码即可,直接返回
return null;
        }
        // 客户端返回
        return React.createElement(props.clientComponent,routeProps);
      }}
    />
  );
}
export default function App() {
  /*
外层 Router 根据环境分开渲染
App 并没有外层的 Router, 仅有 Switch 控制单个渲染,如果访问的不是/、/foo、/baz 等
  路径,则会命中到 RouteWithNotFound 路由上
*/
  return (
    <Switch>
      <Redirect from="/" to="/foo" />
      <Route path="/foo" component={Home} />
      <Route path="/baz" component={Welcome} />
      <RouteWithNotFound clientComponent={NotFind} />
    </Switch>
  );
}
```

上例声明一个 App 组件,其使用了 React Router 提供的 Redirect 与 RouteWithNotFound 自定义组件,分别提供重定向功能与路由定义功能,这主要是为了说明服务端可做重定向和 404 逻辑(Route 为路由端口,Redirect 为重定向组件,将分别在第 6 章和第 8 章介绍)。

在不同的入口将提供不同的 Router,包裹上例中的 App 组件。在 client.js 文件中使用

BrowserRouter 包裹 App 组件提供客户端路由支持：

```js
// client.js 文件，负责客户端渲染
import React from "react";
import ReactDOM from "react-dom";
import { BrowserRouter } from "react-router-dom";
import App from "./App";
// 在客户端使用 BrowserRouter
ReactDOM.hydrate(
  <BrowserRouter>
    <App />
  </BrowserRouter>,
  document.getElementById("app")
);
```

在服务端使用静态路由 StaticRouter 包裹 App 组件提供服务端路由支持：

```js
// server.js 文件，所有执行动作发生在服务端
import express from "express";
import App from "./App";
import ReactDOMServer from "react-dom/server";
const app = express()
app.get('*', (req, res) => {
  // 初始 Context 设置为空，如果对象最终内容不为空，则表明在服务端做了一些处理
  let context = {}
  let html = ReactDOMServer.renderToString(
     // 服务端使用 StaticRouter 将 req.url 作为初始 URL 传入 StaticRouter
     // App 内的 Route 会判断 req.url 是否匹配中某个需要渲染的项
    <StaticRouter location={req.url} context={context}>
      <App/>
    </StaticRouter>
  );
  if (context.url) {
    /*
当 Redirect 组件发生重定向时，context.url 记录了要重定向到的 URL
可使用服务端 304 重定向 context.url 为/foo
    */
    res.redirect(context.url);
    return;
  }
  // 如果上下文中 statusCode 不为空，则说明 Route 对状态码进行了设置，应该返回此状态码
```

```
  // 如果无异常，则发送状态码 200，并将字符发送给客户端
  res.status(context.statusCode || 200).send(html);
});
app.listen(process.env.PORT || 3000);
```

上例使用 ReactDOMServer.renderToString 在服务端获取到需要渲染的 HTML 字符串后，再通过 context.url 判断是否有重定向的情况。其中有两个需要注意的地方：

第一，如果访问根路径/，则 App 中有一个重定向组件 Redirect，在第 8 章将讲到。Redirect 组件会执行 history.push 或 history.replace 方法，本例中 Redirect 组件将根路径/重定向到/foo。正如在 StaticRouter 源码中所看到的，服务端的 history.push 或 history.replace 方法会改变 context 对象的 URL 字段。因而在执行完 ReactDOMServer.renderToString 之后，所有组件的 render 都执行了一遍，context 的内容进行了一次更新，context.url 被更新为/foo。如果在执行完 ReactDOMServer.renderToString 后，上下文对象中 context.url 有值，则表明需要重定向跳转，这时便可以使用服务端的重定向方法并返回。从浏览器端查看的效果是访问/会刷新一次并重定向到/foo。

第二，如果用户初始访问一个没有声明过的路径，如/not-exist，则当应用在服务端执行时，对于 App 中的 Switch 会命中最后一个 RouteWithNotFound 自定义 Route 组件（Switch 类似于 React.Children.only，在本例中起到辅助作用，将在第 8 章进行介绍）。在这个 RouteWithNotFound 逻辑中，会将上下文 context 对象的 statusCode 状态码设置为 404，在 ReactDOMServer.renderToString 渲染完成后，将优先使用此状态码作为 HTTP 状态码进行返回。如果用户访问了/not-exist，由于将 statusCode 状态码设置为 404，因此返回的 HTTP 状态码也将为 404，用户在页面中会看到"404 没有找到"的提示。

另一种情况，如果用户事先访问了/foo 已定义的路径，那么这时可正常看到页面，但此时用户通过某种前端路由方式访问到/not-exist，由于是前端 BrowserRouter 路由，routeProps.staticContext 为空，不执行服务端逻辑。此时由于路由未命中，将渲染传入的自定义 NotFind 组件，给予用户较为友好的提示。这类前端控制的 404 页面将在 8.6.3 节详细介绍。

5.7 相关 Hooks

在第 3 章曾介绍过 React Hooks，通过 Hooks，可复用公共逻辑，减少代码复杂度，也使得复杂组件变得易于理解。本节将介绍与 Router 相关的 3 个自定义 Hook，为后续示例的学习提供基础支持。

5.7.1　useRouterContext

```
// 从 react-router 包中导入上下文
import { __RouterContext as RouterContext } from "react-router";
import * as React from "react";
import { useContext } from "react";
export default function() {
 const routerContext = useContext(RouterContext);
 const { history, location, match } = routerContext;
}
```

React Router 导出了__RouterContext 上下文，在 React v16.3 及以上版本中，__RouterContext 由 React.createContext 原生接口创建，因而可引入并使用 React.useContext（需 React v16.8 及以上版本）获得组件树中距离当前组件最近的 RouterContext 的值（相关特性可查看 3.1 节、3.2.6 节），从 RouterContext 中可获得 history、location 和 match 等，在后面会经常使用这个方法获取 RouterContext。本书如无特别说明，RouterContext 与从 react-router 包中导入的__RouterContext 等价。

5.7.2　useHistory

除了可从 props.history 中获取到 history，在 React Hooks 中如果希望使用 history 对象，那么也可使用 React Router 提供的 useHistory 自定义 Hook（React Router v5.1.2 及以上版本）：

```
import { useHistory } from "react-router"
export default function() {
 let history = useHistory()
 function handleClick() {
   history.push("/home")
 }
 return (
   <button type="button" onClick={handleClick}>
     Go home
   </button>
 )
}
```

useHistory 返回 history 对象，可在函数组件中使用。useHistory 不要求组件在组件树中的层级位置，任意位置都可以引入 history。

在 React Router v5.1.2 之前的版本中，也可使用 useContext 从 5.7.1 节的 __RouterContext 中获取 history：

```
import { __RouterContext } from "react-router";
import * as React from "react";
import { useContext } from "react";
export default function() {
  const { history } = useContext(__RouterContext);
}
```

这与 React Router v5.1.2 源码获取 history 的方式一致：

```
// React Router useHistory 实现
export function useHistory() {
  return useContext(Context).history;
}
```

5.7.3　useLocation

使用 useLocation 返回 location 对象，location 对象代表了当前的路由路径，可从中获得 pathname 和 search 等信息。与 5.7.2 节的 useHistory 类似，useLocation 也由 React Router 提供（React Router v5.1.2 及以上版本），源码实现如下：

```
export function useLocation() {
  // 同样可以通过上下文获取
  // const { location } = useContext(__RouterContext);
  return useContext(Context).location;
}
```

基于 useLocation，还可自定义 useSearch 和 useHash 等自定义 Hook：

```
export function useSearch() {
  const location = useLocation();
  return location.search;
}
export function useHash() {
  const location = useLocation();
  return location.hash;
}
```

5.8 小结

Router 作为 React Router 应用的导航命令传递者，一般在应用的顶层树中使用，并为后代的 Route 提供支持，其使用了 React 的 Context 特性进行跨组件的数据传递。在现代浏览器中，推荐使用 BrowserRouter，因为其具有较好的持久化特性。HashRouter 一般在旧式浏览器中使用，其使用浏览器的 hash 部分，且不需要服务端配置，如果应用中没有持久化 state 等需求，则可使用 HashRouter。对于 React Native 应用，可使用 NativeRouter，其接口与其他 Router 的接口一致。除了客户端方面的支持，React Router 还为服务端渲染提供了静态路由 StaticRouter，通过引入 StaticRouter，可在服务端获得类似客户端中的导航能力。

若希望获得路由上下文或者使用 React Hooks，那么 React Router 提供了__RouterContext 及 useHistory、useLocation 等自定义 Hook。利用这些能力，开发者可便捷地开发 React Router 应用。

参考文献

[1] https://github.com/ReactTraining/react-router.

[2] https://reactnative.cn/.

第 6 章 Route

本章将系统介绍 React Router 的路由端口 Route。Route 是 React Router 体系中极为重要的一个模块，在 React Router 中起到"引路"的作用。

6.1 Route 是什么

一般认为 Route 定义了路由路径匹配成功后的一系列动作。Route 是一个 React 组件，其 props 属性定义了路由应该如何匹配路由路径，以及定义了路由匹配成功后应该渲染的部分。

在 React Router v4.x 之后，Route 的底层设计已经从静态配置的方式演化为动态渲染的方式，即一切都是组件。这带来的是动态化的好处，可以在运行时改变路由属性，适配复杂场景，如根据屏幕尺寸动态改变路由，或者在运行时移除路由或注入路由等。

6.2 Route 的两个基本要素

Route 有两个基本要素，分别是 path 和组件渲染方式（如通过 component 属性渲染）。其中 path 用来定义应该匹配什么样的浏览器路径，component 用来定义匹配路径成功后渲染哪个组件。这两个组件属性构成了一个 Route 最基本的要素。总结其本质为：匹配进而渲染。在 4.4 节中曾介绍过，Route 写好 path 和 component 两个属性之后，Route 的定义就结束了。

6.2.1 Route 的第一个要素：path

对于 React Router 的 path 匹配，按书写格式可分为 3 种方式，分别为基本路径格式的字符串、正则表达式形式的字符串和综合形式的字符串。其写法符合 Express.js[①]的路径声明风格。

对于基本路径格式的字符串路径写法，即常用的 location.pathname，可使用 URL 定义中的 pathname 部分，需注意有些特殊字符需要转义。对于正则表达式形式的字符串路径写法，可在 path 中加入正则表达式规则，对外部路径进行正则表达式匹配，以获取到正则表达式匹配到的各类结果。React Router 同时也规定了在路由中命名参数的写法，只需要遵循一定的格式即可。对于综合形式的字符串路径写法，可以在正则表达式形式的字符串路径写法的基础之上，结合基本路径格式的字符串路径写法，结合转义等，得到约束性较强的路径匹配。

1. path-to-regexp

Route 自身没有实现路径匹配的逻辑，而是使用了一个成熟的第三方库——path-to-regexp 做路径匹配，这个库在 Node.js 框架中被广泛使用。在 Route 的 path 定义中，所写的路由 path 需要被 path-to-regexp 识别为一个符合其定义规则的合法路径。从本质上讲，path-to-regexp 做的是路径的正则表达式匹配，由于其提供了方便的路径定义规则，在内部做了从路径到正则表达式的转换，因此使开发者不用关心复杂的正则表达式就能写出一个灵活性高的 Route 路径。以 React Router 使用的 path-to-regexp v1.7 为例，其 typescript 定义的路径 path 为：

```
export type Path = string | RegExp | Array<string | RegExp>;
```

path-to-regexp 中的路径为字符串与正则表达式或者两者的混合数组。React Router 进一步做了限制，将 path 的类型限定为 string|string[]，不可传入正则表达式类型 RegExp，减去了正则表达式的类型传入，但可以使用正则表达式的字符串形式进行正则匹配，并且对于数组类型的 path，得到的是"或"的关系，即数组中任意一个路径 path 匹配成功后，都将使得整个数组 path 匹配成功，如：

```
<Route path={["/users/info", "/profile/info"]} component={User} />
```

只要任意一个/users/info 或/profile/info 命中，路由即可匹配成功。因此，介绍单个 string 类型的 path 便可同时了解 path 的两类定义规则。

[①] Express.js 是一个基于 Node.js 平台的极简、灵活的 Web 应用开发框架，它提供一系列强大的特性，可以创建各种 Web 应用和移动设备应用。Express.js 不对 Node.js 已有的特性进行二次抽象，只是在它之上扩展了 Web 应用所需的基本功能。

2. 3 种合法的 path

1）基本路径

基本路径即诸如路径/a、/b、/user、/info 等浏览器简单路径，其特点是与 location.pathname 一致，没有对路径附加额外的限定规则。基本路径有着开发简便、便于维护等特点，也是日常开发中使用最多的路径匹配方式，如：

```
// 匹配路径/user/info
<Route path="/user/info" component={A} />
// 匹配路径/user/name
<Route path="/user/name" component={B} />
// 匹配路径/c
<Route path="/c" component={C} />
```

在基本路径匹配中，一般情况下，写上对应的路由路径即可。但需注意，对于闭合的左右圆括号"（""）"，以及"："" * ""\"符号，需要进行转义，否则将被视为正则表达式规则的一部分，从而与期望的结果相悖。原因在于这些符号在 path 中有特殊的表达含义，如不转义将会被识别为特殊的匹配规则。关于具体特殊含义，将在综合路径部分进行阐释。如对于以下路由路径的匹配：

```
// 完全匹配/ab*c 路径，需要转义*
<Route path="/ab\*c" component={ABCD} />
// 完全匹配/a(b)c 路径，需要转义"("，使圆括号不封闭
<Route path="/a\(b)c" component={ABCD} />
// 完全匹配/ab:c 路径，需要转义：
<Route path="/ab\:c" component={ABCD} />
```

在为变量赋值转义字符时，字符串需要写两个反斜线符号，如 const path ="/ab*c"。这是由于 JavaScript 的字符串在遇到一个反斜线符号和一个一般字符时，会将这个反斜线符号作为转义符号进行处理，如变量字符串 a\cd 其实是字符 acd。如果需要保留反斜线符号，则需要再对反斜线符号进行一次转义。

除了上述的特殊情况且在没有额外限制规则的情况下，正则表达式中的"."" + ""?""[""]""-"等均可不用转义，这些字符不会被认为是正则表达式中的符号，其在路径匹配中仅被作为字符串处理，如：

```
// 匹配/a/b+c 路径
<Route path="/a/b+c" component={ABC} />
```

```
// 匹配路径 /readme.md，"."不为正则表达式中的符号，不会匹配 /readme-md、/readme:md 等
<Route path="/readme.md" component={ReadMe} />
```

"+"和"."字符将作为常规路径字符，不作为正则表达式字符，不计入 path-to-regexp 库的路由匹配的规则中。但在路径匹配 path 附加有额外限制规则的情况下，部分字符也有特殊用处，这将在 6.3 节进行阐述。

在一般情况下，如果一个 Route 没有传入 path 参数，则这个 Route 将会永远匹配成功，如下所示：

```
// 无 path 参数的 Route
<Route component={About} />
```

但也有诸如父子 Route 的例外情况，在此种情况下，无路径 path 的 Route 将会匹配失败，在 6.3.1 节会有所阐述。

上例中的 About 组件将永远得到渲染，这在某些场景下会很有用，在 8.6.3 节的实例部分会进行介绍。

2）正则表达式路径

① 基本正则表达式匹配。

除了基本的路径匹配，在 path-to-regexp 中，还有一种通过正则表达式字符串来匹配路径的写法。基本正则表达式匹配也叫无参数匹配，原因在于，对于正则表达式，都会匹配得到一个对象，表明该正则表达式匹配到的结果。由于没有在匹配路径中标明参数，所以得到的结果也仅有匹配命中得到的字符串及匹配命中下标。

对于正则表达式匹配场景，需要在正则表达式路径部分加上封闭的圆括号，如：

```
// 正则表达式 new RegExp('b?')
<Route path="/(b?)" component={ReadMe} />
```

这里对字符 b 做正则表达式处理，匹配 0 个和 1 个 b 字符，即能匹配到/和/b。同理：

- /(b+)能匹配 1 个和多个 b 字符，如/b、/bb、/bbbb 等。
- /(b*)能匹配 0 个和多个 b 字符，如/、/b、/bbbb 等。

若要匹配/a 路径后数字的情况：

```
<Route path="/a/(\d+)" component={ReadMe} />
```

这时正则表达式匹配会生效，会匹配诸如/a/123、/a/456 的路由路径，同时会通过 props 得到具体的匹配结果，得到的匹配变量会用下标代替：

```
// /a/(\d+) 匹配 /a/123
```

```
{
    0:'123'
}
```

② 参数化正则-命名参数匹配。

在 path-to-regexp 中，有一种参数化写法，参数本身不参与匹配，参数化写法能匹配到除 "/" "."（视路由前缀形式决定）外的任意字符。通过提供一个冒号加参数的形式，如/:name，来进行参数化路径匹配，也称为命名参数匹配。通过命名参数匹配，可以在路径中获得相关的业务信息，这种路径的设计方式在 RESTful 中比较常用。当路由到/abc 时，/:name 会成功匹配并且得到如下对象：

```
// 匹配结果
{
name:'abc',
}
```

在 6.3.1 节会介绍通过 props.match.params 获取命名参数匹配结果，如以 name 作为属性，获取匹配到的 URL 值。

在上面的例子中，没有看到用圆括号所表示的正则表达式，其实在 path-to-regexp 内部，以 ":" 开头的路径也是一种正则表达式匹配，原因在于 path-to-regexp 对这类参数化匹配做了特殊处理，其分为两步：

第一步，先匹配 ":" 后面的字符，按照正则表达式/\w+/的匹配原则匹配，包含大小写字母、数字和下画线，得到要匹配的变量名。这个变量名将只做保存用，不参与真实路径匹配。

第二步，运用正则表达式/[^/]+?/进行匹配，即匹配除 "/" 外的所有字符。所以/:name 的路由 path 其实等价于/([^\/]+?),之后会将匹配到的真实路径部分作为属性 name 的值保存下来，name 部分不参与真实路径匹配，如：

```
// 命名参数匹配，name 部分不参与真实路径匹配
<Route path="/:name" component={A} />
// 等价于如下正则表达式匹配情形，且以 name 作为属性保存下来
<Route path="/([^\/]+?)" component={A} />
```

事实上，可以替换参数化匹配时默认的正则表达式/[^/]+?/，写入自行定义的正则表达式，如：

```
// 为命名参数定义正则表达式
<Route path="/:name(\w+)" component={A} />
```

上例中限制了 name 参数需要满足/\w+/正则表达式规则，因而如/a.b.c 类似路径将会匹配失

败。若不加/\w+/限制,以默认规则进行匹配,则将得到如下匹配结果:

```
// 以默认规则进行匹配能匹配"."
{
  name:'a.b.c'
}
```

而加了/\w+/限制,由于/\w/不能匹配".",所以会匹配失败。

③ 正则修饰符匹配。

对于参数化,以及非参数化的正则表达式,可在圆括号结尾处加修饰符"+""*""?"辅助正则表达式匹配,这类在圆括号外的第一个特殊字符称为正则修饰符,如:

```
// 匹配 /c、/c/c、/c/c/c 等
<Route path="/(c)+" component={ABCD} />
// 匹配 、/c 等
<Route path="/(c)?" component={ABCD} />
// 匹配 、/c、/c/c、/c/c/c 等
<Route path="/(c)*" component={ABCD} />
```

与在正则表达式中写上"+""*"相比,在圆括号结尾处进行修饰可以有重复路由前缀的能力,而不是重复正则表达式本身,如:

```
// 匹配/1/2/3 而不是 /123
<Route path="/(\d)+" component={ABCD} />
```

在一些情况下,"+""*""?"也会仅修饰圆括号中的正则表达式,不包括路由前缀。

此外,在以":"开头匹配参数化路径时,也可以结合正则表达式符号"*""+""?"做正则表达式扩展,如:

- /:foo/:bar?,表示 bar 参数为 0 个或 1 个,即/abc 或/abc/1 都将命中。
- /:foo/:bar+,表示 bar 参数为 1 个或多个,即/abc/1 或/abc/1/2 都将命中这个参数路径,命中产生的变量分别如下。

```
// 多个命名参数匹配
{
  foo: 'abc',
  bar: '1',
  }
// 带正则修饰符匹配结果
  {
    foo: 'abc',
```

```
    bar: '1/2',
  }
```

/:foo/:bar*表示 bar 参数为 0 个或多个，与/:foo/:bar+类似，但是它还能匹配命中/abc 的情形，即不匹配 bar 参数。如需要限定/:foo/:bar 中的 bar 为数字，且为多个数字路径的情况，则可以将/:foo/:bar 写为/:foo/:bar(\d+)+，为 bar 这个参数引入"/\d+/"的正则表达式限制，/abc/efg 将会匹配失败，只有/abc/1 和/abc/1/21 等，当第二个参数是数字时才能匹配成功。

值得一提的是，路由的前缀符号除了常用的"/"，path-to-regexp 也允许使用"."，如.a.b 和.:foo.:bar，这在匹配如文件后缀时能起到作用，如：

```
// 匹配 index.d.ts 与 index.ts
<Route path="/index.(d)?.ts" component={ABCD} />
```

3）综合路径

除了第一类基本路径、第二类正则表达式路径，还存在一些路径，如/route.:ext(json|xml|html)+、/(apple-)?icon-:res(\d+).png 和/icon-\(\(*(\d)+\))等，它们为第一类路径与第二类路径的组合形式，且存在一些额外的判定规则，或存在一些需要转义的部分，此种组合情况需要结合路径上下文进行判断。

前面介绍过"*"的转义形式为*，以及跟在圆括号后的形式()*，或者为正则表达式圆括号中的"*"，如(\d*)。除了以上情况，在其他情况下，路径中的"*"字符被视为特殊字符，将匹配 0 个或多个任意字符，注意也包括路径分隔符"/"，例如以下路由路径：

```
// * 匹配任意字符
<Route path="/ab*cd" component={ReadMe} />
```

将匹配/abcd、/abbcd、/abccd、/ab/acd 等。由于"*"将匹配 0 个或多个任意字符，因此也会得到非命名参数，path-to-regexp 内部会将"*"符号转换成/.*/正则表达式，如有以下路由路径：

```
<Route path="/a/*" component={ReadMe} />
```

那么对于路径/a/123，123 将满足/.*/的匹配，也会得到下标为 0、值为 123 的匹配对象。要注意"*"会匹配任意字符，包括路径分隔符"/"。如果有路径/a/b/c，那么对于/a/*的匹配，得到的非参数化对象将会是：

```
// /a/b/c 路径匹配 /a/*
{
    0:'b/c'
}
```

前面介绍过，正则表达式修饰符"+""*"，写在圆括号结尾或者命名参数后可以有重复路由前缀的能力，而不是重复正则表达式本身。但事实上这里需要满足一个前提条件：匹配到路由前缀。

React Router 的路由前缀可以为"/"".", 而能匹配到路由前缀的情况是路由前缀"/""."后紧跟特殊字符，如"("（需要闭合）、":"特殊字符，所以仅在/()、/:key 前缀符号后面跟着闭合的圆括号或":"时，才能拥有重复路由前缀的能力。如有如下匹配路径：

```
/*
    仅匹配/a2、/a12 等，由于圆括号前的路由前缀符号不是"/"或"."，这里是 a，所以
不能匹配/a/1
*/
<Route path="/a(\d)+" component={ABCD} />
```

上例中的修饰符"+"将不会包括路由前缀"/"，此种情况等同于路径/a(\d+)，这也符合书写这个路径时的设想。对于正则表达式修饰符，也有匹配到的路由前缀为空的情况：

```
// 第二个圆括号前的路由前缀符号为"+"，"+"作为修饰符修饰圆括号中的内容，不参与真实匹配
<Route path="/(\d)+(\w)+" component={ABCD} />
```

在此种情况下，因为/\w/没能匹配中路由前缀"."或"/"，所以此时正则表达式修饰符的修饰效果与写在括号内效果一致。

对于"?""*"这样可以匹配空分组的正则表达式修饰符，在匹配空分组的时候，也有一些特定规则，如：

```
// 匹配部分连带前缀匹配空字符串（忽略前缀）、/、/1、/1/2 等
<Route path="/(\d)*" component={ABCD} />
```

即可匹配空字符串（在应用中一般不会出现空字符路径）、/、/1、/1/2、/1/2/3，前缀"/"受正则表达式修饰符"*"的影响。

但是若路由前缀存在且修饰符为"?"或"*"，并且其后跟有非前缀字符的非空字符，如：

```
// * 后紧接字符 a，将匹配/abc、/1abc 等，不能匹配空字符串
<Route path="/(\d)*abc" component={ABCD} />
// ? 后接字符 i，正则表达式修饰符"?"将不影响前缀"/"，能匹配/icon-10.png 等
<Route path="/(apple-)?icon-:res(\d+).png" component={ABCD} />
```

则正则表达式修饰符在匹配空分组的时候，影响的范围不包括前缀，上述将匹配/abc、/1abc、/1/2abc、/1/2/3abc 等。注意路由能匹配到/abc，而不是省略了前缀"/"的 abc，这也与书写时期望的相一致。

要注意，有重复路由前缀能力的正则表达式修饰符"+""*"在非空匹配时，将包含前缀"/"，而不是仅修饰正则表达式部分：

```
/* *后为空, *影响前缀, 将匹配/a/b、/a/b/、/a/b/1、/a/b/1/2 等 */
<Route path="/a/b/(\d)*" component={ABCD} />
/*
    *后紧接字符 c, *不影响前缀, 前缀不能忽略, 将匹配/a/b/c 或 /a/b/1c、/a/b/1/2c
*/
<Route path="/a/b/(\d)*c" component={ABCD} />
/* 匹配 /route.json、/route.json.json、/route.xml.json.html 等 */
<Route path="/route.:ext(json|xml|html)+" component={ABCD} />
```

如果需要匹配()本身, 则需要在所有能形成闭合圆括号处进行转义, 如要匹配/((a))路径, Route 的 path 至少应有两处转义, 如应为/\(\(a)), 使得没有可以闭合的圆括号:

```
// 转义"("或者")", 需要不出现闭合的圆括号
<Route path="/\(\(a))" component={ABCD} />
// 转义"("与"*", 匹配/icon-((*10))
<Route path="/icon-\(\(\*(\d)+))" component={ABCD} />
```

对于有多个圆括号的情况, 最后一个能闭合的圆括号为匹配到的结果, 这时最后一个闭合的圆括号不参与匹配, 无论其余圆括号转义与否, 此时都将作为正常字符串参与路由匹配。

在用圆括号包裹后, 圆括号内部的字符串将作为正则表达式进行处理, 且圆括号内部将不能再使用圆括号, 因为会有新的封闭的圆括号出现; 也不能进行转义, 因为转义符号在圆括号中将执行常规字符的作用, 但是圆括号内的圆括号会再次进行转义, 字符"\("在 path-to-regexp 内部由于"("将转义为"\(", 将会转换得到/\\(/的正则表达式, "("将作为分组特殊符号存在, 与预期的正则表达式/\(/不相符。

以一个例子进行说明, 如要匹配一个以 a 或者 b 开头, 圆括号内是数字的路由, 如/a(1)。在一个正则表达式内表示出该匹配, 当写为/([ab](\d+))时, 因为正则表达式只有字符串从左到右遍历时最先闭合的左右圆括号部分生效, 所以该表达式仅能匹配出/([ab]1, 仅有\d+被作为正则表达式看待, 其余部分都将作为字符串。若对最内部的圆括号进行转义/([ab]\(\d+\)), 这时虽然闭合部分是整个圆括号中的内容, 但是由于"("会再次进行转义, 且前一个字符也是转义字符"\", 因而在 path-to-regexp 内部转换时会得到/\\(/。当这样处理时, 会造成意想不到的结果, "\"由于被转义, 会被当作要匹配的字符, 想要转义当作普通字符处理的圆括号"("会被当作正则表达式的特殊分组符号不参与字符匹配, 因而无法匹配/a(1)。对于此种情况, 可以把正则表达式拆成两部分, 把一个圆括号内的内容拆成多个进行解决:

```
// 匹配/a(1)、/b(12)等
<Route path="/([ab])((\d+))" component={ABCD} />
```

前面圆括号中的正则表达式匹配 a 或 b，接着匹配一个左括号（转义与否均可），再进行数字的正则表达式匹配，最后匹配一个右括号。这里的单个左右括号字符转义与否均可，原因在于库会进行一次判断，对于没转义的，path-to-regexp 内部会进行一次转义；对于转义过的，则不做处理。

6.2.2 Route 的第二个要素：组件渲染方式

Route 的一大职责是渲染组件，将组件渲染方式按路由控制范围从小到大排序，分别是通过 component 属性渲染，获得路由渲染的控制力度较小；通过 render 属性渲染，能获得一定的自由度；通过 children 属性渲染，能获得完全开放的自由度。

1. 通过 component 属性渲染

每一次路径变化，都会进行一次渲染。如果路径匹配成功，就将 component 传入的组件通过 React.createElement 方式进行创建，并且注入 match、location、history 变量，进行 Route 的渲染工作。如果路径匹配失败，Route 则会返回 null，曾渲染过的 component 组件将会被销毁，这时页面 DOM 会消失。当路径为/acd 时，渲染 ReadMe 组件：

```
// 当路径为/acd 时，渲染 ReadMe 组件
<Route path="/acd" component={ReadMe} />
```

当路径不匹配时，通过 component 属性渲染的组件将不会被渲染。若上例从路径/acb 跳转到/abcc，那么通过 component 属性渲染的组件将会执行 componentWillUnmount 生命周期，进而销毁。在一些情况下，这可能是开发者所期望的，因为再次回到这个页面的时候，组件会重新挂载，执行 componentDidMount 生命周期，这样得以再次进行初始化操作。但是在一些页面切换情况下，可能不希望组件被销毁，不希望 DOM 消失。例如用户上传过的大量图片已经通过 createURL 保存在了内存中且展示了出来，如果销毁 DOM，则需要利用变量保存这些 URL，以便页面再次进入时重新渲染。或者对于一些编辑程度很强的页面，如果让 DOM 消失，比起保留 DOM，恢复 DOM 的代价要大得多，这就需要考虑其他办法。

2. 通过 render 属性渲染

通过 render 属性渲染，开发者能自行接管 React.createElement 的行为，在匹配成功后，可以通过渲染插槽的形式渲染对应的组件。相对于通过 component 属性渲染，通过 render 属性渲染能控制传入组件中的 props 参数。

```
<Route
  path="/a/b/c"
  render={(props)=>{
    <Component {...props}/>
  }}
/>
```

如果需要更改传入组件中的 props 参数,则可以使用此种方式。由于此种渲染方式能获得路由中的 history,因此可以包装 history.push、history.replace 等方法,再传递给业务组件。如可以在跳转过程中加入 Hooks 等:

```
// 通过 render 属性渲染方式改变传入的 history
<Route
path="/a/b/c"
render={(({history, ...rest}) => {
    const newHistoryPush = (...args) => {
        if (typeof args[0] === 'string' && args[0] !== '/a/b/c') {
            //这时就是离开/a/b/c的时候
        }
        history.push(...args);
    };
    const newHistory = {
        push: newHistoryPush,
        ...history
    };
    return <Component {...props} history={newHistory} />;
}}
/>
```

当调用 history.push(),离开当前路由时,会执行 newHistoryPush,可以在函数中插入想要的逻辑。或者在某些情况下需要屏蔽一些路由细节,可以修改 this.props.location 后再传入对应的业务组件中,这时对应的业务侧只能感知到修改之后的地址信息。也可以在业务组件外层接入父组件,例如业务权限组件等:

```
// 父组件 Permission 提供控制
<Route
path="/home"
render={props => (
    <Permission code={123}>
        <Component {...props} />
```

```
      </Permission>
  )}
/>
```

上例中的 Component 组件可以受到父组件 Permission 的控制。

对于 React 来说，通过 render 属性渲染也叫作 Render Props 模式。Render Props 模式指的是使用一个值为函数的 props 来传递需要动态渲染的组件。如在下面的代码中可以看到，DataProvider 组件包含了所有跟状态相关的代码，而 User 组件则可以是一个单纯的展示型组件，这样一来 DataProvider 组件就可以单独复用了：

```
import User from 'components/user'
class DataProvider extends React.Component {
  constructor(props) {
    super(props);
    this.state = { target: 'tom' };
  }

  render() {
    return (
      <div>
        {this.props.render(this.state)}
      </div>
    )
  }
}

<DataProvider render={data => (
  <User target={data.target} />
)}/>
```

3. 通过 children 属性渲染

通过 children 属性可自由地控制渲染，此种渲染方式拥有最高的自由度。若在 children 函数中返回组件，则组件将无条件被渲染，如下所示：

```
// 无条件渲染 Component 组件
<Route
  children={({match, location, history}) => (
```

```
    <Component />
  )}
/>
```

或者

```
// 无条件渲染 Component 组件
<Route>
  {({match, location, history}) => (
    <Component />
  )}
</Route>
```

children 函数在 Route 被渲染时将无条件被调用。通过 children 属性进行渲染时，Route 注入了 3 个参数：match、location、history，渲染的工作完全交给了开发者。

同时，由于 React Router 把该注入的 3 个参数交给了开发者，因此提供了很大的发挥空间，开发者可以利用注入的 3 个参数自行定义渲染逻辑。

通过使用 children 属性，可以将一些逻辑改变为路由逻辑。如已经编写了一个通过 visible 属性控制弹窗可见的组件：

```
<Modal visible={this.state.visible}/>
```

如果希望通过路由路径控制弹窗，并且不改变原有路由路径结构，那么常用的办法是增加一个 query 参数，通过读取此参数实现控制。若不加 query 参数而希望通过路由路径控制弹窗弹出关闭，则可在原路径上继续加一级路径，同时使用 Route 的 children 属性渲染这个弹窗组件，并为 Route 设置一个匹配 path。通过路由形式，可以通过改变 URL 控制弹窗展示，如下所示：

```
// 用 match 参数来做 Modal 弹窗控制
<Route
  path={`${props.match.params.path}/modal-A`}
  children={({match}) => {
    <Modal visible={match} />;
  }}
/>
```

当前的组件获取到路由传入的 props，能获取到 match 参数，若当前匹配渲染的路径 path 是 /a/b，即 props.match.params.path 的值为 /a/b，则只需把路径改成 /a/b/modal-A，弹窗便会弹出，且 /a/b 对应的原页面也存在并且不受路由变化的影响。这利用了 Route 为动态路由的特性，路由 path 不是静态配置好的。

通过 children 属性进行渲染，还可以有更多的控制，如页面间动画过渡，也可以使用 children 属性进行渲染。在 6.7 节的 Route 实战案例中，也会常用到这个属性，例如把 Route 改造为保留 DOM 的 CacheRoute。

6.3　Route 传入组件的 3 个参数

6.3.1　match

1．match 对象各属性

match 对象一般从 props.match 中获得，使用 render 和 children 渲染方式也可获得 match 对象，同时在上下文中也能获取 match 对象。match 对象具有如下 4 个属性：params、isExact、path 和 url。

1）params

params 是一个对象，记录了参数化配置时路由匹配出的键值对，如：

```
<Route path="/user/:name" component={User} />
```

如果路由路径为/user/tom，则能从 User 的 props 中得到的 props.match.params 为：

```
{
    name:'tom'
}
```

如果命名参数加上圆括号，以正则表达式进行限制：

```
<Route path="/ages/:age(\d+)" component={Age} />
```

则只有第二个参数是数字时才能匹配上，如/ages/18。注意，如果第二个参数是非数字，如/ages/jim，则 match 为 null，视为匹配失败。

2）isExact

isExact 表示 Route 是否完全匹配，如果路由注册的 path 是/a/b，那么在 Route 的 exact 是 false 时且路由在路径为/a/b/c 的情况下也能匹配中；但是，如果 isExact 为 true，就只能进行精确匹配，路由 path 为/a/b 将无法匹配/a/b/c。

3）path

path 为匹配到的路由 Route 的 path 属性，此属性通常用于嵌套路由中，作为父路由部分使用。

由于从 React Router v4.x 开始，Route 的 path 属性需要完全路径形式，不能再像 React Router v3.x 中一样支持部分路径，如父路由为/a/b，则子路由只需为/c 就可匹配/a/b/c。若使用最新版本的 React Router 需要满足此种场景，则可以使用父路由传入的 match 对象中的 path 字段，这个字段记录了父路由 Route 命中后，父路由 Route 的 path 部分，此刻可以使用 props.match.path 作为子路由完全路径中的父路由部分，如：

```
// 利用匹配中的 path 做二次匹配
<>
  {props.match && (
    <Route
      path={`${props.match.params.path}/c`}
      component={SubComponent}
    />
  )}
</>
```

props.match 可能为 null，即为父路由没有匹配中的情况，使用时需注意。

在上例中，Route 是动态渲染的，将使用拼接路径进行一次路由匹配。当路由匹配命中后，对应的 SubComponent 就会进行实例化，得到渲染。

4）url

url 为真实的 URL 匹配命中部分，可以在嵌套导航场景中使用，如 match.params.url 可作为某二级 Link 的导航前缀。如果希望导航时在当前匹配路径下增加一级导航路径，但是不知道当前路径的具体值，则可以使用 match.params 中的 url，如：

```
// 利用匹配中的 params
<Link to=`${props.match.params.url}/c` />
```

如果当前匹配路径为/a/b，若 React Router 运行在浏览器中，那么将会使浏览器导航到/a/b/c 下。

2. 没有命中路由时 match 的情况

考虑一种情况，当父 Route 利用 children 进行无条件渲染时，子组件传入一个无 path 的子 Route 组件，这时子 Route 组件的渲染情况。

```
// when location.pathname is /matches
<Route
  path="/does-not-match"
  children={({match}) => {
```

```
  return (
    <Route
      render={(({match: pathlessMatch}) => {
        // can not render
        return <div>123</div>;
      }}
    />
  );
}}
/>
```

在一般情况下，在没有写 path 的情况下，Route 将会视为匹配成功，将会渲染对应组件而不关心路由路径，这样的 Route 也称为 pathless Route。但是，在上述情况中，嵌套的 Route 在事实上是无法成功匹配的，即在路由路径为/matches（与/does-not-match 不匹配）的情况下，并不会渲染得到 123。这是因为对于无路径的 Route，如果嵌套使用在一个父路由 Route 中，其 match 对象将会继承父路由 Route 的 match 对象，所以上述例子中的父 match 对象为 null，继承得到的 pathlessMatch 也为 null。由于无 path 路由使用的是 render 渲染，在 match 不为空时才能得到调用，所以组件自然无法渲染出来。

在这种情况下，要想子路由得到渲染，需要写明子路由的 path 属性，而不能写一个没有 path 属性的子路由，或者这时候子路由使用 children 属性进行无条件渲染。

6.3.2 location

location 一共有 5 个参数，如下所示：

```
{
  key: 'ac3df4',
  pathname: '/a',
  search: '?b=c',
  hash: '#something',
  state: {}
}
```

其中，key 在 hash 路由下不存在，为 null，在浏览器路由中为一个随机值，在第 2 章曾经描述过 key 的作用及产生原理。location，顾名思义，即位置，pathname 为 URL 路径的 path 部分，如 https://www.github.com/a/b/c 中的/a/b/c 部分。search 属性为路径的 query 部分，注意这里包括

"?"。hash 属性为路径的 hash 部分，注意包括 hashTag "#" 符号。此外还有一个 state 对象，即作为 pushState/replaceState 接口中可以持久化存储的状态。在浏览器路由中，可以使用 state 对象作为信息传递载体。如果当前的路由模式为 hash 路由，情况则稍有不同。当路由模式为 hash 时，路径部分将从 "#" 后开始计算，也就是忽略了 window.location.pathname，只使用浏览器 hash 部分计算路由，如 https://www.github.com/a/b/c?d=1#/user，location 中的 pathname 为/user，从 "#" 后开始计算路由，在第 2 章曾有过介绍。

对于浏览器 query，React Router 并没有解析成一个对象，只能从路由中得到?some=search-string 这样的字符串，不过可以借用 URLSearchParams 对象轻松实现，如：

```
// 解析 query
let params = new URLSearchParams(location.search);
```

Route 可以自行传递 location，在一般情况下不需要手动给 Route 传入 location，Route 会从上下文中获取 location 对象。但在有些场景中，如 Route 置于某个父容器中，而这个父容器希望缓存 Route 的状态，则可以把当前拿到的 location 传入 Route。父容器接管 Route 后，由于父容器可以控制子容器的渲染，因此其可以先将 location 状态缓存一段时间再进行处理。如果不将 location 传入 Route，那么 Route 将从 context 中获取到最新的 location，绕过了父容器，这时也就无法缓存了，这个场景在 8.6.1 节会进行介绍。

6.3.3　history

history 在第 2 章介绍过。history 的 location 属性在每次路由路径变化时都会进行更新。在业务开发中，如果需进行以下前后渲染判断，则不推荐从 history 中获取地址信息：

```
class Comp extends React.Component {
  componentDidUpdate(prevProps) {
    /* location 每次渲染都不同，locationChanged 将会永远为 true，表明每次渲染得
到的 location 是不同的，每次渲染都有自身的 location */
    const locationChanged =
      this.props.location !== prevProps.location;
    /* 不推荐这样使用，locationChanged 将永远为 false，由于 history 是单例的，在常
驻内存中，通过 history 获取的 location 都是同一个引用 */
    const locationChanged =
      this.props.history.location !==
      prevProps.history.location;
  }
```

```
}
<Route component={Comp} />;
```

history 对象自创建完成后就驻留在内存中，为一个单例的对象，其不由 React Router 维护。history 由 history 库创建后便由 React Router 交由开发者，开发者可以随意改变单例的 history 对象，对象是可变的。对于上例，自 history 对象创建完毕，React Router 就将其透传到业务中，所以每次渲染得到的 history 都是相同的，因而 history 的各属性也没有渲染上的区别，在做前后渲染判断逻辑时其上的属性不应该受到信任。而不像 props.location，location 每次都会重新创建，这确保了每次渲染得到的 props.location 都不同。建议在 React Router 中获取 location 时使用 Route 的 props.location 来替代 history.location。

这里的 history 为第 2 章中的 browserHistory、hashHistory 和 memoryHistory 中的一种，通过 Router 组件透传到 Route 中，接口与第 2 章中介绍的一致。

6.4 Route 的其他配置

6.4.1 location

Route 允许自行传入 location 进行匹配，而不是使用上下文中的 location。传入的 location 的 pathname 可以与当前的 pathname 不相同，这可在某些场景中发挥作用，如路由动画等。关于路由动画的示例，可查看 8.6.1 节。

6.4.2 exact

exact 表明路由配置为完全匹配，如有路由：

```
<Route path='/base' component={Comp} />;
// /base/path 路径匹配成功
```

使用上述/base 路由配置，若当前的路径为/base/path，当 exact 为 false 时，上述路由会匹配成功；若设置 exact 为 true，则表明路由为完全匹配模式：

```
<Route exact path='/base' component={Comp} />;
// /base/path 路径匹配失败
```

此时/base/path 路径会匹配失败，只有/base 或/base/路径才能匹配成功。在表 6-1 中，列出了不同 exact 配置与不同 Route 的 path 属性在不同匹配路径下的匹配结果。

表 6-1　不同 exact 配置与不同 Route 的 path 属性在不同匹配路径下的匹配结果

exact 配置	外部的路径地址	path 属性	Route 的路由匹配结果
true	/a	/a	true
true	/a/	/a	true
true	/a/b	/a	false
false	/a/b	/a	true
true	/a	/a/	true
true	/a/	/a/	true
true	/a/b	/a/	false
false	/a/b	/a/	true

从上面的例子中可以看到，<Route path="/a">将匹配部分路径，如其能模糊匹配/a、/a/b、/a/b/c、/a/b/c/d 等。如果希望完全匹配，则需要设置 exact 为 true：

```
<Route exact path='/a' component={Comp} />;
```

在嵌套路由或父子路由场景中，exact 的取值需要注意，这关系到子路由的匹配，将在 6.7.1 节嵌套 Route 示例部分进行介绍。

6.4.3　strict

strict 表明设置路由为严格模式。当 strict 为 true 时，需要严格按照 Route 的 path 中最后的分隔线进行匹配，如有以下设置了 strict 的 Route 配置：

```
<Route strict path='/base/' component={Comp} />;
// /base 路径无法匹配成功
```

在上例中，由于 Route 是 strict 模式，需要/base/路径才能匹配成功。表 6-2 列出了不同路径下不同路由 path 在 strict 设置下的匹配结果。

表 6-2　不同路径下不同路由 path 在 strict 设置下的匹配结果

strict 配置	外部的路径地址	path 属性	Route 的路由匹配结果
true	/a	/a	true
true	/a/	/a	true
true	/a/b	/a	true
true	/a	/a/	false
false	/a	/a/	true

续表

strict 配置	外部的路径地址	path 属性	Route 的路由匹配结果
true	/a/	/a/	true
true	/a/b	/a/	true

从表 6-2 中可看到，strict 能确保匹配如/base/路径的严格匹配。如果需要绝对完全路径匹配，如 Route 的 path 为/a/b，且在路由的匹配上也仅希望匹配/a/b 路径，而不匹配/a/b/、/a/b/c 等路径，则需要将 strict 与 exact 都设置为 true，才能保证仅有/a/b 路径能成功匹配此路由，而/a/b/与/a/b/c 都不能成功匹配。

```
<Route exact strict path='/a/b' component={Comp} />;
```

6.4.4 sensitive

sensitive 用于设置 Route 的路由匹配是否对大小写敏感，默认为不敏感。例如有以下路由：

```
<Route path='/base' component={Comp} />;
// /BASE 路径匹配成功
```

能匹配/BASE 路径。若设置了 sensitive：

```
<Route sensitive path='/base' component={Comp} />;
// /BASE 路径匹配不成功，仅能匹配/base 路径
```

则仅有/base 路径能匹配成功，/Base 和/BASE 等路径都将匹配失败。

6.5 Route 源码解析

6.5.1 上下文的更新

在第 5 章中曾介绍过 Router 提供了初始上下文，其中包含初始的 match 信息、history 对象、初始 location 等。且 Router 在内部监听了 history 的 location 变化，并在每次 location 变化后为 Provider 提供更新后的 location，供 Consumer 消费。而一些位于较上层的 Route 组件，依赖于在组件树中的位置，正是 Router 中 Provider 的消费者。与 Router 不同的是，Route 既是 Context 树中 Provider 的消费者，同时又是新的上下文提供者，如图 6-1 所示。

Context 树中最顶层组件为 Router，用以提供初始的上下文。在 Router 的上下文被后代 Route

消费后，后代 Route 为后继的 Route 提供了新的上下文，后继 Route 将消费此新的上下文。在 Route 源码实现上，源码在 RouterContext.Consumer 消费上下文后又提供了新的 RouterContext.Provider：

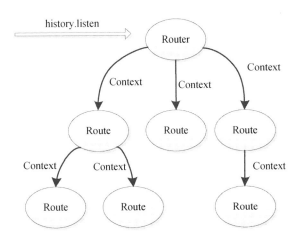

图 6-1　Route 在 Context 树中提供新的上下文

```
class Route extends React.Component {
  render() {
    return (
      <RouterContext.Consumer>
        {context => {
          // 确保使用了上下文
          invariant(context, "You should not use <Route> outside a <Router>");
          // 可从 props 中设置 Route 的 location，一般不设置，默认从上下文中获取
          const location = this.props.location || context.location;
          // 计算该路由是否命中，可从 Switch 组件已计算值中获取，或者使用
          // matchPath 进行计算，或者继承上下文中的 match
          const match = this.props.computedMatch
            ? this.props.computedMatch // <Switch> already computed the match for us
            : this.props.path
            ? matchPath(location.pathname, this.props)
            : context.match;

          // 构造新的上下文
          const props = { ...context, location, match };
```

```
      let { children, component, render } = this.props;
      // Preact 兼容
      if (Array.isArray(children) && children.length === 0) {
        children = null;
      }

      return (
        // 提供新的上下文
        <RouterContext.Provider value={props}>
          {props.match
            ? children
              // 在匹配成功进行渲染时 children 函数渲染优先级最高
              ? typeof children === "function"
                ? children(props)
                : children
              // component 渲染优先级次之
              : component
              ? React.createElement(component, props)
              // render 渲染优先级最低
              : render
              ? render(props)
              : null
            // 在匹配失败时调用 children 函数渲染
            : typeof children === "function"
            ? children(props)
            : null}
        </RouterContext.Provider>
      );
    }}
  </RouterContext.Consumer>
    );
  }
}
export default Route;
```

通过 Context 的接管与更新的传递模式，React Router 的上下文就在这样的 Context 树中被更新了。

6.5.2 运行流程

如图 6-2 所示，在源码实现上，Route 将所有逻辑写在 RouterContext.Consumer 中，目的是获得 Router 或 Route 传递的上下文，通过对上下文对象进行读取，能获知上级 Route 的 match 对象、location 地址和 history 对象。在代码运行时，Route 首先判断 context 对象的有无，如果没有 context 对象，则检查不通过，Route 将报错，无法运行。其次，Route 需要外界的 location 地址信息，获取 location 的目的是获得当前的 pathname，即当前的路由路径，为路由匹配做好准备。Route 会尝试使用外部传入的 location，如果传入 location，则优先使用传入的 location，这一步可为 Route 提供可控制的 location，在 8.6 节中会看到传入 location 的相关示例。注意，如果自行传入了 location，由于上下文的更新传递，传入的 location 会同时传递给 Route 的子级 Route 以供消费。

而如果未将 location 传入 Route，则 Route 将从上下文中获取 location，注意上下文由组件树中距离最近的上级 Route 或者顶层 Router 提供，如果距离最近的上级 Route 改变过 location，由于上下文的传递，则 context.location 也是修改之后的 location。之后 Route 将进行路由匹配的计算，将确认是否命中路径。为了确认 Route 是否命中路径，首先 Route 会判断是否传递了 computedMatch，这个属性为内部维护的 prop，由 Switch 组件提供，在 8.1 节会进行 Switch 的介绍。如果传递了 computedMatch，则表明此 Route 的命中情况已被 Switch 计算过，可直接使用此传入 prop。如果 Route 不是作为 Switch 的子组件，则不会接收此属性。之后，Route 会判断是否有 path 属性，如果有此属性，则会调用 matchPath 进行路径匹配计

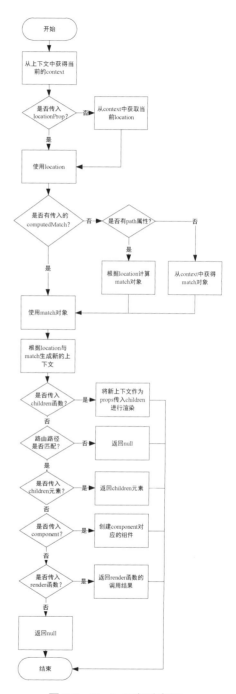

图 6-2　Route 运行流程图

算。关于 matchPath，将在 8.5 节进行介绍。matchPath 判断当前 pathname 与 Route 的 path 属性是否匹配，如果匹配，则返回 match 对象；否则 match 为 null。这里的 match 即 6.3.1 节中介绍过的 props.match。

如果未将 path 传入 Route，则此种情况称 Route 为 pathless 的 Route，或者称为无路径 Route。无路径 Route 的 match 会使用 context.match，即上下文中的 match。如果无路径 Route 的上下文的 context 由第 5 章中所讲的 Router 提供，那么无路径 Route 将视为路由匹配命中。在 5.2.2 节曾介绍过，Router 路由器提供的初始上下文中的 match 不为 null，match.path 为 "/"，此时无路径 Route 即可消费到此 match 对象。以 6.3.1 节介绍的父 Route 未命中，而子 Route 为 pathless 的示例为例：

```
<Route
  path="/does-not-match"
  children={({match}) => {
    return (
      <Route
        render={({match: pathlessMatch}) => {
          // 不会渲染
        }}
      />
    );
  }}
/>
```

在上例中，由于子 Route 没有传入 path 属性，此时的 match 就会使用 context.match 父 Route 中的 match 对象，而父 Route 并没有命中，其 match 为 null，因此子 Route 的 match 也为 null。因而子 Route 也不会渲染。

在明确了 match 对象与 location 对象后，渲染前的准备工作就完成了，得到了 Route 所能提供的 props，其内容为 6.3 节介绍的 3 个属性变量。接下来可以开始渲染工作，在 6.2.2 节曾介绍过组件的 3 种渲染方式，分别是通过 children 属性渲染、通过 component 属性渲染和通过 render 属性渲染。如果传入了 children 函数，则调用 children 函数将 props 传入，得到返回的结果，并直接渲染 children 函数的返回结果。可看到 Route 在路由命中与未命中时，都对 children 函数无条件调用，而如果 children 不为函数，这时的 children 可能为合法的 React 组件元素，则仅在路由命中时渲染 children。如果没有 children，而是传入了 component 或者 render，则将会根据路由命中与否进行组件创建或者调用。component 的优先级比 render 高，它们的渲染优先级依次为 children、component、render。

RouterContext.Provider 又重新把计算得到的 props 作为上下文向下传递。这一步非常重要，为组件树中的所有子 Route 提供新的上下文信息。

6.6 相关 Hooks

6.6.1 useRouteMatch

useRouteMatch 由 React Router 在其 5.1.2 版本中提供，其功能为获取某路径的路由匹配情况或者获取上下文中的 match 命中情况：

```
import { useRouteMatch } from "react-router-dom";
function Component() {
  // 判断/foo/baz 是否命中当前 location 地址
  let match = useRouteMatch("/foo/baz");
  // 获取上下文中的 match 命中情况
  //  let match = useRouteMatch();
  return <div />;
}
```

其实现如下：

```
export function useRouteMatch(path:string) {
  return path
    ? matchPath(useLocation().pathname, path)
    : useContext(Context).match;
}
```

useRouteMatch 可接收 path 作为参数，如果传入了 path，则会调用 useLocation 获取当前的 URL 路径，判断 path 与当前路径是否匹配。如果没有传入 path，则会使用上下文中的 match，此 match 可以为 Router 中提供的 match 或者组件树中距离最近一级 Route 提供的 match。

6.6.2 useParams

useParams 同样由 React Router 在其 5.1.2 版本中提供。useParams 将获取上下文中的命名参数匹配结果：

```
<Route path="/user/:id">
```

```
      <User />
</Route>
/* 函数组件 */
function User() {
  let { id } = useParams()
  return <div>用户 id 为{id}</div>
}
```

其源码也较为简单:

```
export function useParams() {
  const match = useContext(Context).match;
  return match ? match.params : {};
}
```

使用 useContext(Context) 获得上下文中的 match。当路由未命中时，match 为 null。对于 match 为 null 的情况，这里返回了空对象，表示无命名参数的匹配结果。

6.7　Route 实战案例

本节将基于 Route 的基础知识，为读者提供实战案例，帮助读者加深对 Route 的理解和使用。

6.7.1　嵌套 Route

在一些开发场景中，Route 组件的子组件或后代组件往往也存在 Route 组件，这些 Route 组件在结构上构成了树状结构。或者从 Route 的声明结构上来说，Route 间相互嵌套，其层级为嵌套层级。

一般来说，这类结构的 Route 称为嵌套 Route，如以下示例:

```
function Foo(prop) {
  return (
    <>
      {/* prop.match.path 由父级传入，此处为/foo */}
      {/* 真实路径为/foo/:id */}
      <Route path={`${prop.match.path}/:id`}>
        {/* 渲染 detail 详情组件 */}
        <FooDetail />
```

```
      </Route>
      {/* 渲染拼接的 list 路径 */}
      <Route path={`${prop.match.path}/list`}>
        <ul>{/* …… */}</ul>
      </Route>
    </>
  );
}
// 父级 Route
<Route path="/foo" component={Foo}/>
```

对于 path 为/foo，渲染 Foo 组件的 Route，由于 Foo 组件中也使用了 Route，因此可称此路由为父路由。在 Foo 组件中，同样使用 Route 组件，由于当/foo 命中时，会将命中结果作为 props 传入，因此可从 props.match 中读取到父级命中信息。可利用此特性，读取 props.match.path 上一级 Route 传入的命中结果，并使用上一级 Route 传入的命中结果作为本级 Route 的路由前缀。

上例中 Route 间的关系仅为父子组件，可读取父组件传入子组件的 props.match.path 并获取父级路径信息。但若希望某个嵌套层级很深的组件也能使用此嵌套路由，且不希望将 props 逐级传递，则可使用 RouterContext 上下文跨组件特性以获取组件树中距离本级组件最近的 Route 的匹配信息：

```
import { __RouterContext } from 'react-router';
// 某层级很深的组件
function SomeDeepComponent(prop) {
  // match 由组件树中距离本级组件最近的 Route 提供
  const { match } = React.useContext(__RouterContext);
  return (
    <>
{/* 拼接路由 */}
      <Route path={`${match.path}/somepath`}>
        {/* 渲染某层级很深的组件 */}
      </Route>
    </>
  );
}
<Route path="/foo" component={Foo} />;
```

通过路由上下文的方式，开发者即使不为子组件传入 props，也能获取到 match 对象。但需要注意的是，需要在 React v16.8 及以上版本中才可以使用函数组件的 Hook 方法 React.useContext。

如果不用 Hook 方法 React.useContext，那么也可使用 RouterContext.Consumer 获得路由上下文对象，这里读者可自行实现。

6.7.2 相对路径 Route

在 React Router v3.x 中，由于 React Router 默认使用了父子路由的形式进行路由声明，因此路由路径的声明无须书写全路径，如以下示例：

```
<Route path="/home" component={AppComponent}>
   <Route path="/dashboard" component={Dashboard}></Route>
 </Route>
```

Dashboard 组件将在路径/home/dashboard 下渲染。而在 React Router v4.x 之后，路由的性质发生了变化，要想 Route 能正确匹配路径，如以上示例中 Route 的路径声明，则需要写为完全路径：

```
<Route path="/home" component={AppComponent}>
   <Route path="/home/dashboard" component={Dashboard}></Route>
 </Route>
```

事实上，在一些开发场景中，路由的声明并不希望写为全路径的形式。比如在某一个层级较深的组件内使用 Route，并不希望先到外层组件中获知路由路径，再写到这个层级较深的 Route 中；也不希望写为 6.7.1 节中的路由拼接形式，需要通过 props.match.url 拼接上 path。为了解决上述问题，可以改造 Route，封装一个相对路径 Route 组件以满足需求：

```
<RelativeRoute path={"/b"} component={Component} />
```

如 RelativeRoute 的上级 Route 路径为/a，在某组件内使用以上组件，将会得到路径为/a/b 的等效 Route，Component 组件将在/a/b 路径下渲染：

```
<Route path={"/a/b"} component={Component} />
```

通过路由上下文，可实现 RelativeRoute：

```
import { RouteProps, Route, __RouterContext } from "react-router";
import * as React from "react";
import { useContext } from "react";
export default function(props: RouteProps) {
  // 获取上级 Route 匹配信息
  const { match } = useContext(__RouterContext);
  let path;
```

```
  if (match) {
    path = `${match.path}${props.path}`;
  }
  // 如果父级 Route 未匹配，则 path 为 undefined
  // 此时本级 Route 为 pathless Route，将使用父级 Route 匹配结果
  return <Route {...props} path={path} />;
}
```

上例中使用了函数组件，并使用 useContext(__RouterContext)获得路由的 Context。Context 中保存了组件层级中距离最近的一个 Provider 所提供的值，这正是组件树中的父级 Route 或者某个祖先 Route。当这个上级 Route 匹配命中时，Context 中的 match 不为 null，可从 match.path 中获取上级 Route 声明的 path 值，这时可以拼接上级 path 与本级 path（示例中忽略了字符串的处理，如拼接"/"字符），得到对应的完全路径。这个路径又可作为一个父路径向下传递。

若上级路由没有命中成功，match 为 null，则此时本级子路由应完全依附于父路由的匹配结果，在一般情况下不能出现父路由未命中，而子路由又命中的情况。这时，可以为本级 Route 传入 undefined 的 path。在 6.5 节中曾介绍过 Route 源码的运行流程，当 Route 不传入 path 或者传入的 path 为 undefined 时，判断 Route 的命中与否将以 context.match 为依据。示例中若上级 Route 的 match 为 null，则本级 Route 的 match 使用 context.match，也同样为 null，这时子路由将匹配失败，这与上级 Route 的匹配结果是一致的。

6.7.3 重定向 Route

在 React Router v3.x 中，React Router 提供了 IndexRedirect 组件，用于为父路由提供默认跳转。IndexRedirect 组件的作用是在父路由命中的情况下，将父路由重定向到一个新的路由地址，如：

```
<Route path="/" component={App}>
  <IndexRedirect to="/welcome" />
  <Route path="welcome" component={Welcome} />
  <Route path="about" component={About} />
</Route>
```

上例在命中路径为 / 的父路由后，渲染 App，同时 IndexRedirect 组件会将地址导航到 /welcome 路径下，此时 React Router 将渲染 App 与 Welcome 父子组件。

使用 IndexRedirect 组件，外层父路由 Route 一般提供框架组件，而 IndexRedirect 组件的 to

属性所指向的地址，一般也定义了 Route，所对应的路由一般提供了具体的页面组件。在开发应用时，IndexRedirect 组件为父路由提供重定向支持，这给定义路由带来了便利。

在升级到 React Router v4.x 之后，IndexRedirect 被取消了。若希望实现类似效果，则可基于 Route 的动态特性扩展 Route，使其获得重定向能力：

```
import { Route, RouteProps, Redirect } from "react-router";
interface RoutePropsExt extends RouteProps {
  // 在原有 Route 属性的基础上，增加命中重定向属性
  indexRedirect?: string;
}
export default function(props: RoutePropsExt) {
  const { component: Component, indexRedirect } = props;
  return (
    <Route
      // 一般情况下，作为父路由
      exact={false}
      {...omit(props, ["component", "children"])}
      render={renderProps => {
        return (
          <>
            {/*在传入了 indexRedirect 且当前路径绝对匹配时，提供重定向导航*/}
            {indexRedirect && renderProps.match.isExact && (
              // Redirect 将在第 8 章介绍，这里可理解为跳转到 to 路径
              <Redirect
                to={`${processPath(renderProps.match.url)}/${processPath(
                  indexRedirect
                )}`}
              />
            )}
            <Component {...renderProps}>{props.children}</Component>
          </>
        );
      }}
    />
  );
}
```

在实现上，通过封装 Route 提供函数组件。组件的 props 除了继承 Route 的 props，还允许传

入 indexRedirect 字符串，作为重定向地址。在 Route 路径的选择上，外部的 path 等属性透传给本级 Route，不对 path 进行额外处理。在渲染模式上，通过读取 props.component 获取到需要渲染的组件，并使用 Route 的 render 属性进行渲染。与此同时，在 render 函数中判断路由匹配情况，实现重定向逻辑。在通常情况下，此组件作为父 Route，为路径匹配提供重定向服务。为了保证子路由的渲染，Route 的 exact 需设置为 false。如果 exact 为 true，则 Route 在完全匹配路径时才渲染组件，这与期望不符。比如父路由路径为/且 exact 为 true，在路径为/home 时，父路由将匹配失败（父路由路径/不完全匹配/home），从而父路由所在的路由树都得不到渲染，这与父子路由设计有出入。

上例中为了使得 Route 获得重定向逻辑，利用匹配结果 renderProps.match.isExact 来判断是否完全匹配中了当前的路由。如果匹配结果是完全匹配并且传入了 indexRedirect 字符串，则可以实例化一个 Redirect 组件（将在第 8 章进行介绍），用于重定向到具体的路径。而在路由匹配成功但 renderProps.match.isExact 为 false（不完全匹配）时，由于 render 函数正常触发，组件依然会得到渲染，而重定向逻辑不会触发。

考虑到组件通常作为父 Route，在重定向路径选取上，使用拼合路径，重定向路径中包括了 Route 命中结果 routeProps.match.url。这样拼合后的 Redirect 路径 to 为 routeProps.match.url 字符串与 indexRedirect 字符串的拼接。由于重定向地址包含当前 Route 的匹配结果，这样保证了在重定向时路径始终包含原命中路径，当前 Route 组件及渲染的 Component 组件不会由于导航到其他路径而被销毁，同时也能保证父路由对子路由的渲染。

封装后的 Route 用法如下：

```
const Layout = props => props.children;
/* Router 提供支持 */
<Router history={history}>
  <RouteWithIndexRedirect
    path="/"
    indexRedirect="/list-A"
    component={Layout /*简单的 Layout*/}
  >
    {/* 与 Route 的行为一致 */}
    <Route path="/list-A" component={A} />
    <Route path="/list-B" component={B} />
  </RouteWithIndexRedirect>
</Router>
```

在上例中，当匹配中路径/时，会渲染 component 指明的 Layout 组件，这里的 Layout 仅为简单的渲染 children 子元素，之后会重定向到/list-A 路径。此时会渲染 Layout、A 两个组件。通过 RouteWithIndexRedirect 的引入，无论是访问路径/，还是访问路径/list-A，得到的结果都相同。

6.7.4 默认子组件 Route

与 IndexRedirect 为父路由提供默认重定向类似，在 React Router 第三版时，React Router 提供了 IndexRoute 作为父路由的补充路由。IndexRoute 是一类补充路由，附属于父路由而存在。IndexRoute 可渲染父路由下额外的组件，其不能拥有 path 属性，其 component 属性所提供的组件用于给父路由提供完全命中时额外子组件的渲染，如：

```
<Router>
  <Route path="/" component={App}>
    <IndexRoute component={Home}/>
    <Route path="accounts" component={Accounts}/>
    <Route path="statements" component={Statements}/>
  </Route>
</Router>
```

上例中/路径会渲染 App 与 Home 组件。这是由于 IndexRoute 为父路由提供了附属组件支持。

通过 IndexRoute 默认子路由形式，父级路由路径不仅提供了外层框架组件，也在满足父级路由绝对匹配的情况下获得了 IndexRoute 的额外组件支持。使用 IndexRoute，由于 React Router 帮助开发者做了组件的组合，将便于应用开发时页面的组件化设计。

而在 React Router 第四版之后，React Router 动态路由有着更强的灵活性，其将原有的 IndexRoute 组件的组合特性移除，并将其交给了开发者，由开发者自由组合组件。若希望在第四版的 React Router 中也能使用类似的补充组件能力，则可封装带默认子组件的 RouteWithIndexComponent：

```
import { Route, RouteProps } from "react-router";
import * as React from "react";
import { omit } from "lodash";
interface RoutePropsExt extends RouteProps {
  // 默认子组件属性
  indexComponent?: React.ComponentType;
}
```

```
export default function(props: RoutePropsExt) {
  return (
    <Route
      {...omit(props, ["component", "children"])}
      exact={false}
      render={renderProps => {
        const { component: Component, indexComponent: IndexComponent } = props;
        return (
          <Component {...renderProps}>
            {/*渲染默认子组件*/}
            {IndexComponent && renderProps.match.isExact && (
              <IndexComponent {...renderProps} />
            )}
            {props.children}
          </Component>
        );
      }}
    />
  );
}
```

使用 RouteWithIndexComponent 可以同时渲染两个组件，一般用于父子路由，外层父路由为页面框架或组件框架，子路由为具体的业务组件。

在实现上，通过封装原生 Route，提供函数组件。组件的 props 除了继承 Route 的 props，还允许传入 indexComponent 组件，作为路由的默认子组件。在路径匹配上，外部的 path 等属性透传给本级 Route，不对 path 进行额外处理。在渲染模式上，通过读取 props.component 获取到需要渲染的组件，并使用 Route 的 render 属性进行渲染。与此同时，也在 render 函数中判断路由匹配情况，决定是否需要渲染 component 的子组件。

注意，默认子组件 Route 与 6.7.3 节的重定向 Route 类似，其 exact 同样为 false，这是为了保证父路由的模糊匹配，相关内容在 6.7.3 节已有介绍。

与第三版 Route 类似，在 RouteWithIndexComponent 组件正常匹配渲染时，将渲染传入的 component 组件及其 children 部分 props.children（通常为子组件元素）。而在路由完全匹配（renderProps.match.isExact 为 true）的场景中，除了上述渲染，还渲染传入的 indexComponent 组件，作为 component 的子组件。

这样就在路径绝对匹配时渲染了传入的 component 组件与其子组件 indexComponent。如果路径不是完全匹配，即 renderProps.match.isExact 为 false，则 indexComponent 组件不会渲染，只会渲染 component 组件及 props.children，与预期一致。

RouteWithIndexComponent 的使用方式为：

```
<RouteWithIndexComponent
        path="/"
        indexComponent={A}
        component={Layout /*props => props.children*/}
    >
        <Route path="/list-B" component={B} />
</RouteWithIndexComponent>
```

在上例声明的路由下，若当前路径为"/"，则在命中路由"/"后，会渲染 Layout 组件与 Layout 组件的子组件 A。而如果路由路径为/list-B，由于不满足 indexComponent 组件渲染所需的路径绝对匹配条件（/list-B 不绝对匹配/），A 组件不会渲染，RouteWithIndexComponent 只会渲染 component 组件及 props.chilren，即 Layout 组件与 B 组件。

6.7.5　缓存 Route

在 React Router 中，一般情况下原生 Route 所负责渲染的组件在命中路由时进行挂载，而在导航时离开，路由未命中时组件将被销毁，分别对应了组件的 componentDidMount 与 componentWillUnmount 生命周期。由于导航离开、组件被销毁，组件的状态也被清理，而当再次访问路由地址时，又重新执行了挂载逻辑，所有页面状态，包括 DOM 节点都重新初始化。

而在某些业务场景中，开发者希望导航前后页面的状态得到缓存，而不希望页面状态在导航离开后被框架回收，不希望 componentWillUnmount 生命周期被执行。

在 Vue 或 Angular 中，有对应的 keep-live 与路由复用策略可实现页面缓存，而在 React 或 React Router 中，都没有提供相关的缓存能力。要想实现页面缓存，在 React 应用中通常有以下两种解决方案。

第一种方案：状态存在内存中。比如使用 Redux、Mobx 或自定义内存变量，在页面离开前，将页面内用户产生的数据存储在内存中，将页面"快照"下来，并销毁页面的 DOM 节点。这样虽然页面的 DOM 节点被销毁，但是在重新挂载页面的时候，可从内存中将各节点通过"快照"进行恢复。这种方案的优点主要是路由切换 DOM 节点销毁，DOM 节点数量可控，而其缺点也同样明显，将引入额外开发量，无论是在业务上还是在框架上，都将引入额外的开发成本。

第二种方案：不销毁 DOM 节点，对其进行缓存。这种方案的优点主要是可以降低开发成本。虽然页面的 DOM 节点会保留，但是由于减少了页面离开时的"快照"过程，业务开发的流程侵入性较小，并不需要在页面销毁时将页面状态保存到内存中。通过缓存 DOM 节点，一来无须框架层面的额外开发，二来也不会侵入业务代码，仅需要对路由的行为做一些改变。但这类方案也同样存在缺点，由于一般通过 CSS 方式缓存 DOM 隐藏页面，因此页面的 DOM 节点数量会增多。但若控制得当，DOM 的节点数量也在可控范围内。

开发者需在开发成本与性能间进行考量，如业务情况允许可选择第二种方案，因为不仅能有效地降低开发成本，而且在控制得当的情况下，对性能的影响也较小。本节将介绍第二种方案（CacheRoute）的实现。此方案实现的核心在于使用 Route 的 children 属性进行可控渲染。CacheRoute 除了拥有原始 Route 的能力，还基于 Route 提供了几类扩展能力：路由组件状态缓存、remount 机制和渲染优化。

1. 路由组件状态缓存

通过 6.2.2 节介绍的 Route 的 children 属性，组件的渲染逻辑可由开发者控制。利用此特性，可在 match 为 null（路由未命中）时设置 style 隐藏组件，实现 CSS 隐藏式页面切换：

```
<Route
  children={({ match, location, history }) => (
    <div style={{ display: match ? "block" : "none" }}>
      <Component />
    </div>
  )}
/>;
```

2. remount 机制

remount 机制指的是 React 组件进行一次 unmount 销毁，清除 DOM，再重新挂载 DOM 的机制。在 React 中可有多种方案实现，常见的做法是通过给同一个 React 组件设置不同的 key，利用 React 的 diff 机制。如果 key 相同，则组件执行的是更新周期；如果发现 key 不同，则会销毁原组件，重新挂载新的 key 对应的组件。如在每个组件初始化时生成了 key 值，在需要重新挂载时再为该组件重新生成另一个 key 值，便可实现组件的重新挂载。

在组件实现思路上，以父级组件提供能力支持。组件初始化时使用 React.useRef 声明一个随机的 key，当组件收到 remount 信号，即当 props.shouldRemountComponent 为 true 时，更改 key

的值并保存到 Ref，并对组件使用此更新后的 key。这样当组件应用 key 时，由于 keyRef.current 的变化会使得组件被销毁并重新渲染：

```
import * as React from "react";
export default function Remount(props) {
  // 初始化 key
  const keyRef = React.useRef(Math.random() + Date.now());
  if (props.shouldRemountComponent) {
    // 更新 key
    keyRef.current = Math.random() + Date.now();
  }
  return React.cloneElement(React.Children.only(props.children), {
    key: keyRef.current
  });
}
```

3. 渲染优化

对于通过 CSS 缓存的组件，在路由切换过程中，无论路由命中与否，由于组件没有被销毁，组件都会执行更新的生命周期，页面在未命中路由时也会调用 render 函数，这在路由未命中时是没有必要的。引入记忆渲染的能力，在类组件中，可通过父组件或高阶组件的 shouldComponentUpdate 判断是否有必要渲染子组件；或者对于函数组件，使用 React.memo 记录渲染结果。在 3.4 节中曾介绍过 React.memo，其第一个参数为函数组件，第二个参数为更新判断函数，可根据路由匹配与否选择是否渲染页面。类似于类组件中的 shouldComponentUpdate，使用 React.memo，React 会缓存上一次渲染结果。本节通过判断 nextProps.match 及 nextProps.match.isExact，提供两类记忆化父组件，如果 nextProps.match 或 nextProps.match.isExact 有值，则说明路由匹配命中，应该渲染；反之，则缓存上次渲染结果，组件不进入渲染的生命周期：

```
import * as React from "react";
import { RouteChildrenProps } from "react-router";
interface Props extends RouteChildrenProps {
  children?: any;
}
function MemoChildrenWithRouteMatch(props: Props) {
  return props.children;
}
export default React.memo(MemoChildrenWithRouteMatch, (prvious, nextProps) => {
```

```
  // 不命中就不渲染，仅在 match 有值时才渲染组件
  return !nextProps.match;
});
function MemoChildrenWithRouteExactMatch(props: Props) {
  return props.children;
}
export const MemoChildrenWithRouteMatchExact = React.memo(
  MemoChildrenWithRouteExactMatch,
  // 不命中就不渲染，仅在 match 有值且绝对匹配时才渲染组件
  (prvious, nextProps) => !(nextProps.match && nextProps.match.isExact)
);
```

可将 Route 的匹配结果 routeProps 传入 Cache 组件，并将需要缓存的组件作为 Cache 组件的子组件：

```
<Route
  path={path}
  children={(routeProps: RouteChildrenProps) => {
    <MemoChildrenWithRouteMatch {...routeProps}>
      {/* …… */}
      {/* <Component/> */}
    </MemoChildrenWithRouteMatch>;
  }}
/>;
```

结合以上三方面能力，使用 React Hooks 实现 CacheRoute 如下：

```
import * as React from "react";
import { Route, RouteChildrenProps, RouteProps } from "react-router";
import { omit } from "lodash";
import MemoChildrenWithRouteMatch, {
  MemoChildrenWithRouteMatchExact
} from "./MemoChildrenWithRouteMatch";
import Remount from "./remountComponent";
interface Props {
  forceHide?: boolean;
  shouldReMount?: boolean;
  shouldDestroyDomWhenNotMatch?: boolean;
  shouldMatchExact?: boolean;
}
export default function CacheRoute(props: RouteProps & Props) {
```

```
      const routeHadRenderRef = React.useRef(false);
return (
  <Route
    {...omit(props, "component", "render", "children")}
    children={(routeProps: RouteChildrenProps) => {
      const Component = props.component;
      // 获取 Route 命中结果
      const routeMatch = routeProps.match;
      let match = !!routeMatch;
      if (props.shouldMatchExact) {
        // 使用 isExact 进行判断
        match = routeMatch && routeMatch.isExact;
      }
      if (props.shouldDestroyDomWhenNotMatch) {
        if (!match) routeHadRenderRef.current = false;
        // 按 react-router 包中的 Route 逻辑
        if (props.render) {
          return match && props.render(routeProps);
        }
        return (
          match && Component && React.createElement(Component, routeProps)
        );
      } else {
        const matchStyle = {
          // 隐藏
          display: match && !props.forceHide ? "block" : "none"
        };
        if (match && !routeHadRenderRef.current) {
          // 将渲染标志设置为 true
          routeHadRenderRef.current = true;
        }
        let shouldRender = true;
        if (!match && !routeHadRenderRef.current) {
          shouldRender = false;
        }
        // 选择对应的 Memo
        const MemoCache = props.shouldMatchExact
          ? MemoChildrenWithRouteMatchExact
```

```
            : MemoChildrenWithRouteMatch;
        // CSS 隐藏保留 DOM
        let component;
        if (props.render) {
          component = props.render(routeProps);
        } else {
          component = <Component {...routeProps} />;
        }
        return (
          shouldRender && (
            {/*提供 CSS 属性 */}
            <div style={matchStyle}>
              {/*提供 remount 能力*/}
              <Remount shouldRemountComponent={props.shouldReMount}>
                {/*提供渲染优化*/}
                <MemoCache {...routeProps}>{component}</MemoCache>
              </Remount>
            </div>
          )
        );
      }
    }}
    />
  );
}
```

CacheRoute 的 Props 在原有 RouteProps 的基础上增加了 4 个 props：

```
interface Props {
  shouldReMount?: boolean;
  shouldDestroyDomWhenNotMatch?: boolean;
  forceHide?: boolean;
  shouldMatchExact?: boolean;
}
```

在某些需要清除组件状态的场景中，可传入 shouldReMount。shouldReMount 表示是否需要销毁组件并重新渲染 DOM，即通过该属性实现组件的重新挂载。shouldDestroyDomWhenNotMatch 表明 DOM 的渲染模式为销毁模式。如果传入为 true，则在路由未命中时，原有页面的 DOM 会被销毁，将与原始 Route 行为一致。forceHide 为强制隐藏的属性，如果为 true，则强制 Route 渲染的 DOM 为 CSS 隐藏，无论命中与否（这个属性是为扩展的 Switch 提供的，在 8.6.4 节会进行介绍）。shouldMatchExact 表示判断组件缓存时是全匹配缓存还是模糊匹配缓存，可选择模糊命中缓存组件（不传入）与绝对命中缓存组件（设置为 true），其实现在于 match 与 match.isExact 的区别。使用 CacheRoute 的示例方式为：

```
<CacheRoute path="/baz" component={Baz}>
```

这样渲染的组件不会被销毁，由于 shouldDestroyDomWhenNotMatch 默认为 undefined，CacheRoute 渲染 Baz 组件时根据 RouteProps 的 match 来计算包裹组件的 CSS 样式：

```
const matchStyle = {
  // 隐藏
  display: match && !props.forceHide ? "block" : "none"
};
```

如果 props.forceHide 为 true，则 display 为 none，强制隐藏 CSS。

同时，根据传入的命中模式，选择对应的 Memo 函数：

```
const MemoCache = props.shouldMatchExact
        ? MemoChildrenWithRouteMatchExact
        : MemoChildrenWithRouteMatch;
```

为了兼容原 Route 的渲染模式，CacheRoute 也支持 Render Props 渲染模式：

```
let component;
if (props.render) {
  component = props.render(routeProps);
} else {
  component = <Component {...routeProps} />;
}
```

需要注意，当存在某页面初始加载成功，其余页面路由未命中的情况时，如/a 页面初始加载，其余/c、/d、/e……路径下的路由未命中，如果未经任何逻辑处理，则组件会以 CSS 隐藏方式得到渲染进而挂载。这种情况应该注意，/c、/d 等页面组件不应渲染，应只有路由命中成功之后，才能进行初始渲染及后续对 DOM 进行保留。代码中使用 routeHadRenderRef 来记录此状态：

```
const routeHadRenderRef = React.useRef(false);
```

只有当 match 为 true 时，才会设置 routeHadRenderRef.current 为 true。如果组件是 DOM 销毁，则也应把 routeHadRenderRef.current 设置为 false。

6.7.6　Route 渲染组件的可访问性支持

可访问性（WAI-ARIA）是指无障碍的网页应用技术，主要是为了提升网页的可用性，将网页无障碍化。通过 WAI-ARIA，可以表达 HTML 无法自行表达的语义，它具备比现有的 HTML 元素和属性更完善的表达能力，能让页面中元素的关系和含义更明确，从而使得类似屏幕阅读器的设备能准确识别网页中的内容。以一个示例说明，考虑如下代码片段：

```
<li tabindex="0" class="checkbox" checked>复选框</li>
```

列表项 li 元素作为一种自定义复选框。名为 checkbox 的 CSS 类为元素提供了所需的视觉效果。虽然这适合视力正常的用户，但是对弱视用户来说，屏幕阅读器却不会给予任何指示来说明该元素的用途是作为复选框。

通过使用 WAI-ARIA 属性为元素提供缺少的信息，屏幕阅读器便能正确解读该元素：

```
<li tabindex="0" class="checkbox" role="checkbox" checked aria-checked="true">
  复选框
</li>
```

上例代码中添加了 role 和 aria-checked 属性，将该元素显式地标识为一个复选框，并指定它在默认情况下处于选中状态。屏幕阅读器将把它正确地报告为一个复选框。

应用于 HTML 的 WAI-ARIA 属性由两部分组成：role 和带 aria-前缀的属性。role 属性描述了一个元素的实际作用；带 aria-前缀的属性描述了元素在无障碍交互情境中的具体信息。

对于有障碍的用户，为其提供页面辅助支持，将解决交互的障碍问题。本节通过 React Router 的扩展，在不影响原有页面能力的情况下，增强屏幕阅读器对导航的感知。

由于 React Router 的使用通常为 SPA 应用，页面与页面之间通常仅是不同的 HTML 元素，缺少一些语义标识。这时可使用 role 为 group 的元素包裹页面组件：

```
<div tabindex="-1" role="group" style="outline: none;"></div>
```

或者使用 role 为 page 的元素包裹页面组件：

```
<div tabindex="-1" role="page" style="outline: none;"></div>
```

在无障碍操作时，屏幕阅读器会将当前获得焦点的元素的相关 aria 信息进行报告朗读。但对于 React Router 单页面应用来说，页面间的跳转并不会改变页面的焦点状态，因而这需要人为控制，引入高阶组件：

```
export default function(pageComponent) {
  const Cls = pageComponent;
```

```
    return props => {
      const wrapDiv = React.useRef<HTMLDivElement>();
      function focus() {
        wrapDiv && wrapDiv.current && wrapDiv.current.focus();
      }
      useEffect(() => {
        // 挂载时聚焦
        focus();
      }, []);
      return (
        <div ref={wrapDiv} tabIndex={-1} style={{ outline: "none" }} role=
"group">
          <Cls {...props} />
        </div>
      );
    };
}
/* 引入高阶组件获得 aria 支持 */
const UserWithARIA = withARIA(User);
<Route path="/user" component={UserWithARIA} />;
```

React.useRef 用以保存包裹 div 的 node 信息,可在组件挂载时调用元素原生的 focus 方法,聚焦外层 div 元素,使之获得焦点,从而使得屏幕阅读器获知当前页面的元素情况。如果在 Mac 上使用 voiceOver,则会读出包裹 div 下的 children 元素信息。

6.7.7　query 及命名参数

在第 1 章曾介绍过使用 location.search 获得浏览器 query,并实例化对象 URLSearchParams 解析浏览器 query:

```
import React from "react";
import { BrowserRouter, Link, Route } from "react-router-dom";
export default function() {
  return (
    <BrowserRouter>
      <Route path="/" component={ParamsDemo} />
    </BrowserRouter>
  );
}
```

```
}
function ParamsDemo(props) {
  // 从 props 的 location 中获取 search
  const params = new URLSearchParams(props.location.search);
  return (
    <div>
      <Link to="/user?name=foo">foo</Link>
      <h3>query 中的参数为{params.get("name")}</h3>
    </div>
  );
}
```

对于不支持 URLSearchParams 对象的浏览器，也可使用 query-string 等第三方库解析浏览器 query 部分。

如果希望在函数组件中使用 query，则可自定义 Hook，如 useQuery：

```
import React from "react";
import { __RouterContext } from "react-router";
// 使用第三方库解析
import queryString from "query-string";
// 以 use 开头的自定义 Hook
export const useQuery = () => {
  const { location } = React.useContext(__RouterContext);
  const { search } = location;
  const query = React.useMemo(() => queryString.parse(search), [search]);
  return query;
};
```

这里使用了 React Router 的内部上下文 __RouterContext，并从中获得当前的 location，从 location 中可获取 search 字符串，再使用 React.useMemo 记录 queryString.parse 得到的结果。这里依赖为 search，仅 search 变化时再次调用 queryString.parse，如果 search 不变化，便不需调用 queryString.parse，直接返回上一次调用所得到的值。在 React Router v5.1.2 之后，增加了自定义 Hook。如上例可从 React Router 中引入 useLocation（5.7.3 节）获得 location 对象。

对于命名参数的获取，也可以使用 React Router 所提供的 useParams：

```
<Route path="/user/:id">
    <User />
</Route>
/* 函数组件 */
```

```
function User() {
  let { id } = useParams()
  return <div>用户 id 为{id}</div>
}
```

或者如 6.3.1 节中所介绍的，从 props.match 中获取。

6.7.8　Route 中的代码拆分

对于一个实际应用来说，控制所需加载的 JavaScript 的大小是有必要的，程序应当只加载当前渲染页所需的 JavaScript，这也称为代码拆分，即将所有的代码拆分成多个小包，在用户浏览过程中按需加载。由于 Route 的职责是配置好每个路径的视图，所以在 Route 侧非常适于做代码拆分。在 React Router v3.x 中，通过 getComponent 延迟路径匹配特性支持了组件的异步加载，而在 React Router v4.x 及之后的版本中，路由性质由静态路由变更为动态路由。由于动态路由在运行时才得以确立，其天然地支持了路径的延迟匹配，因此 getComponent 接口被废弃，组件的异步加载使用常规的 component 渲染便可实现。如果依赖于 webpack 等代码拆分工具，那么可以使用动态 import 语法（dynamic import），这类工具会将动态导入的模块进行代码拆分。相对于静态 import，动态 import 作为一个函数，常在按需加载模块时使用，其将返回一个 promise。当该 promise 被 resolve 后，得到的内容即需要 import 的内容：

```
import('/modules/my-module.js')
  .then((module) => {
    // 得到 module
  });
```

通过动态导入，可方便对路由配置异步组件：

```
function getComponentAsync(loader) {
  return props => {
    const [component, setComponent] = React.useState(null);
    React.useEffect(() => {
      loader().then(asyncComponent => {
        setComponent(asyncComponent);
      });
    }, []);
    return component && React.createElement(component, props);
  };
```

```
}
<Route path="/user" component={getComponentAsync(()=>import('./user.jsx'))}/>
```

在以上例子中,要注意异步模块的导入形式为() => import("/someModule")。这是为了延迟 import 的执行,如果在程序入口直接调用 import("/someModule"),那么虽然也能有异步拆分代码的效果,但是在入口即会将模块全部异步加载,这样不会有每个页面分别加载的效果。

也可以使用 react-loadable 等第三方库组件:

```
import Loadable from "react-loadable";
const LoadableUserComponent = Loadable({
            loader: () => import("/user"),
            loading: () => {
              return <div>Loading...</div>;
            }
          });
<Route path="/user" component={LoadableUserComponent}/>
```

在 React v16.6 之后的版本中,可使用 React.suspent 与 React.lazy 进行模块的异步加载工作:

```
import React, { Suspense } from "react";
// 动态加载 User
const LazyUserComponent = React.lazy(() => import('./User'));
function DynamicUser(props) {
  return (
    <Suspense fallback={<div>Loading...</div>}>
      <LazyUserComponent {...props}/>
    </Suspense>
  );
}
<Route path="/user" component={DynamicUser}/>
```

6.8 小结

Route 是 React Router 中的核心组件之一,理解并学会使用 Route 对理解应用的路由结构有重要意义。

本章首先介绍了 Route 在 React Router 中所起的作用及其所扮演的角色,并介绍了 Route 的基本使用方法,且详细地介绍了 Route 的路径匹配,包括基本路径、正则表达式路径、综合路径匹配；以及路由组件的 3 种渲染方式，包括通过 component 属性渲染、通过 render 属性渲染和通过 children 属性渲染。还介绍了 Route 的各类配置，并通过 Route 源码解析，帮助读者加深理解 Route。同时，为读者介绍了 React Router 提供的与 Route 相关的 useRouteMatch 与 useParams 自定义 Hook，并通过嵌套路由 Route、相对路径 Route、CacheRoute、Route 的代码拆分等实战案例，为读者提供了丰富的实战指导。

参考文献

[1]　https://github.com/ReactTraining/react-router.

[2]　https://developers.google.com/web/fundamentals/accessibility/semantics-aria?hl=zh-CN.

第 7 章 Link

Link（导航）作为 React 应用中的重要一环，是应用中不可缺少的部分。无论是通过 history 导航，还是使用组件导航，只要能灵活地掌握不同的导航方法，对应用的开发就极有帮助。本章将介绍 React Router 中的 Link 组件，通过对 Link 组件的学习，可帮助开发者在 Link 开发时更灵活地进行组件式开发设计。

7.1 Link 介绍

7.1.1 Link 的定义及属性

Link 组件提供声明式的导航方式，相比直接通过 history 导航，其将渲染 DOM 元素，具有组件特性。

由于 Link 组件带有 DOM 元素，因此其需要从 react-router-dom 包中引入：

```
import { Link } from "react-router-dom";
```

Link 的 props 定义为：

```
export interface LinkProps<S = H.LocationState>
  extends React.AnchorHTMLAttributes<HTMLAnchorElement> {
  to: H.LocationDescriptor<S>;
  replace?: boolean;
  innerRef?: React.Ref<HTMLAnchorElement>;
}
```

其中 to 定义了 Link 要导航到的具体地址，replace 定义了是否替换历史栈方式进行导航，innerRef 提供了 Link 内部 DOM 元素的引用。一般来说，Link 组件将渲染 a 标签元素。

1. to 属性

LinkProps 的 to 属性在 Link 中定义为 LocationDescriptor，可为字符串或地址描述对象：

```
export interface LocationDescriptorObject<S = LocationState> {
    pathname?: Pathname;
    search?: Search;
    state?: S;
    hash?: Hash;
    key?: LocationKey;
}
export type Path = string;
export type LocationDescriptor<S = LocationState> = Path | LocationDescriptorObject<S>;
```

当 to 接收一个字符串时，可将路由导航到指定路径：

```
<Link to="/courses?sort=name#the-hash" />
```

或者 to 为地址描述对象：

```
<Link
  to={{
    pathname: "/courses",
    search: "?sort=name",
    hash: "#the-hash",
    state: { fromDashboard: true }
  }}
/>
```

相比字符串模式，此写法能传入 state 对象。要注意 state 对象在 hash 路由模式下虽然也可以传递，但因为在 hash 模式下没有 pushState 进行状态保留，将不会有状态持久化特性，相关内容可查看 2.3 节。

随着浏览器历史栈管理能力逐渐普及，未来 hashHistory 有望支持状态的持久化。

此外，在 React Router v5.1 及以上版本中，Link 还支持 location 为 function 的传入：

```
<Link to={location => ({ ...location, pathname: "/courses" })} />
<Link to={location => `${location.pathname}?sort=name`} />
```

这样可以得到当前的地址信息，并可基于当前地址信息进行跳转。

2．replace 属性

当传入的 replace 为 true 时，Link 的底层行为是 replaceState，对应 BrowserRouter 为浏览器原生的 replaceState 接口，对应 HashRouter 为 window.location.replace，只替换 hash 部分。如果是内存路由，则模仿浏览器路由，更改自身维护的栈内容即可。

3．innerRef 属性

innerRef 可以接收一个函数，能获得真实 DOM 的引用。

```
const refCallback = node => {
  // `node` refers to the mounted DOM element or null when unmounted
}
<Link to="/" innerRef={refCallback} />
```

或者在 React v16.3 之后，有更推荐的方式，使用 React.createRef：

```
const anchorRef = React.createRef()
<Link to="/" innerRef={anchorRef} />
```

在 React Router v5.1 之后，由于 Link 源码使用了 React 的 forwardRef 特性，在 React 版本支持的情况下，可传入 Ref 获取内部 DOM 元素的引用：

```
const anchorRef = React.createRef()
<Link to="/" ref={anchorRef} />
```

4．onClick 属性

在不改变 Link 内部渲染 component 的情况下，如果 Link 中传入了 onClick，如：

```
<Link onClick={(event)=>{}} to="/a" innerRef={anchorRef} >
    跳转到/a
</Link>
```

将会捕获到 Link 中元素的单击。也可以在 onClick 事件中调用 event.prventDefault，这样会阻止 Link 的跳转行为。在 Link 的 props 中，没有经过特殊处理的其余 props 都会传递给内部 component 组件，默认为 a 标签，如 title、style 和 className 等。如不需 a 标签发送 Referer 字段，则可设置 rel="noreferrer"。这样当单击导航产生的 HTTP 请求时，就不会带有 Referer 字段。

7.1.2　Link 源码解析

Link 源码在实现上分为两部分，一部分提供导航信息，不关心执行；另一部分负责执行导航操作，不关心导航地址。其也可称为上下文消费组件和导航命令执行组件。如下所示为上下文消费组件，通过它获取上下文的 history 和 location 当前地址信息。上下文消费组件解析了 to 的地址，并判断选取 history 的导航方法，封装好 props 后传入导航命令执行组件中：

```
const Link = forwardRef(
  (
    {
      // 默认传入 LinkAnchor 组件，可自行传入 component
      component = LinkAnchor,
      replace,
      to,
      innerRef,
      ...rest
    },
    forwardedRef
  ) => {
    return (
      <RouterContext.Consumer>
        {context => {
          const { history } = context;
          const location = normalizeToLocation(
            resolveToLocation(to, context.location),
            context.location
          );
          const href = location ? history.createHref(location) : "";
          const props = {
            ...rest,
            href,
            // 传入导航方法
            navigate() {
              const location = resolveToLocation(to, context.location);
              const method = replace ? history.replace : history.push;
              method(location);
```

```
          }
        };
        if (forwardRefShim !== forwardRef) {
          // 在 React v16.3 以上版本中可使用 ref，不用 innerRef
          props.ref = forwardedRef || innerRef;
        } else {
          props.innerRef = innerRef;
        }
        // 将 props 传入导航命令执行组件中
        return React.createElement(component, props);
      }}
    </RouterContext.Consumer>
  );
}
);
export default Link;
```

对于导航真正的执行组件，Link 中使用了内部默认的 LinkAnchor 组件（在 React Router v5.1 之后，可在 Link 中传入 component 覆盖默认的 LinkAnchor 组件）。默认 LinkAnchor 组件用于导航 DOM 元素的渲染，以及导航命令的执行：

```
const LinkAnchor = forwardRef(
  (
    {
      innerRef,
      navigate,
      onClick,
      ...rest
    },
    forwardedRef
  ) => {
    const { target } = rest;
    let props = {
      ...rest,
      onClick: event => {
        try {
          // 执行传入的 onClick 函数
          if (onClick) onClick(event);
        } catch (ex) {
```

```
        event.preventDefault();
        throw ex;
      }

      if (
        // 没有阻止过
        !event.defaultPrevented && // onClick prevented default
        event.button === 0 && // ignore everything but left clicks
        // 如果 target 为 _blank 或其他,则不做处理,将处理权交给浏览器
        (!target || target === "_self") &&
        !isModifiedEvent(event) // ignore clicks with modifier keys
      ) {
        event.preventDefault();
        // 执行传入的导航函数
        navigate();
      }
    }
  };
  if (forwardRefShim !== forwardRef) {
    props.ref = forwardedRef || innerRef;
  } else {
    props.ref = innerRef;
  }
  return <a {...props} />;
 }
);
```

从以上源码中可看到,若 Link 传入了 onClick 函数,则会在 LinkAnchor 的 a 标签元素被单击时执行。在执行导航的判断逻辑上,仅在事件默认行为未被阻止,以及按键、单击、target 等逻辑判断通过后,才会执行传入的 navigate 函数。

7.2 NavLink

7.2.1 带激活态的 Link

NavLink 是 Link 的升级版,除了能支持常规的 Link 能力,NavLink 还能设置当路由匹配成功时 Link 的样式。除了继承 Link 的 props,NavLink 还支持传入以下 props:

```
export interface NavLinkProps<S = H.LocationState> extends LinkProps<S> {
  activeClassName?: string;
  activeStyle?: React.CSSProperties;
  exact?: boolean;
  strict?: boolean;
  isActive?<Params extends { [K in keyof Params]?: string }>(
    match: match<Params>,
    location: H.Location<S>,
  ): boolean;
  location?: H.Location<S>;
}
```

可以看到，NavLink 也有 Route 的 exact 和 strict 属性。事实上，在理解 NavLink 时，可以将其理解为 Route 和 Link 的组合组件。Route 负责匹配传入的 to 路径，并且激活样式；Link 根据传入的 to 属性进行导航。

对于 NavLinkProps 中的 activeClassName 和 activeStyle，可分别设置激活态时应用的类名和 CSS 内联样式。

```
// 当外部路径为/faq 时，将在 a 标签中使用如下 style
<NavLink to="/faq" activeStyle={{ fontWeight: "bold", color: "red" }}/>
// 当外部路径为/faq 时，将对 a 标签应用 active 的 CSS 类
<NavLink to="/faq" activeClassName ="active" />
```

若传入 NavLinkProps 中的 isActive 属性，则将 NavLink 激活与否交给组件外部判断。isActive 接收一个函数，函数参数为(match, location)，可以通过(match, location)加入判断逻辑，如：

```
<NavLink
  to="/user/123 "
  isActive={(match, location) =>        location.pathname.startWith
("/user")}
>
  User
</NavLink>
```

除了基本的导航能力，NavLink 还可以对导航匹配情况进行限制，有 strict、exact 和 location 属性。strict 可类比 Route 的 strict，即限制命中路径需严格匹配。exact、location 与 Route 的 exact、location 类似，目的都是限制 NavLink 在判断激活与否时的规则或自定义外部 location。

为什么 Link 没有 exact、strict 等属性，而 NavLink 有呢？因为 Link 仅是单纯地进行导航，导航路径并没有精确与否的区别；但是由于 NavLink 为 Route 与 Link 的组合组件，所以 Route

的属性部分也应用到 NavLink 上。

对于 NavLink，可设置 aria-current 属性激活态的值，用来辅助屏幕阅读器读出当前页面激活的导航项。与 activeClassName 和 activeStyle 同理，当 NavLink 与当前路由路径匹配成功时，会设置 a 标签的 aria-current 属性，默认设置为 page：

```
<a aria-current="page" />
```

设置了 aria-current，在导航时，屏幕阅读器会读出当前页面元素中 aria-current 为 page 的 Link 中的文本内容，以表明当前导航到的页面。此设置将帮助 Web 应用程序更具可访问性。

7.2.2　转义特殊字符

当 NavLink 导航到一个 URL 路径时，如何判断激活态呢？事实上，NavLink 可以从当前 location 中得到 pathname，然后跟传入的 to 属性进行比对，如果 pathname 和 to 匹配成功，则说明 NavLink 组件是属于激活态的。但是遇到的一个问题是，如果 to 传入的路径带有特殊字符，如：

```
<NavLink to="/user(3)"/>
```

这时如果用 path-to-regrexp 中的 match 方法直接进行匹配，就会匹配失败。试想，当写 Route 路由端口时，也会遇到这个问题。如果想要匹配/user(32)这一类的 path，则 Route 需要写为（在 6.2.1 节曾有介绍）：

```
<Route path="/user\(:id\)"/>
```

可以看到，上例对"（""）"进行了转义。所以对于 NavLink 的匹配状态，即激活态，也要采取这样的策略。当对 NavLink 传入如"/user(3)"时，也需对"（""）"进行转义。在 NavLink 的内部实现上，其对特殊字符进行了转义，相关的正则为([.+*?=^!:${}()[]|\])。在代码实现上，传入的 unescapedPath 作为未转义的 path，会进行 path 转义：

```
const path = unescapedPath.replace(/([.+*?=^!:${}()[\]|/\\])/g, '\\$1');
```

这段代码的意义为：对于路径中包含的"." "+" "*" "?" "=" "^" "!" ":" "$" "{" "}" "(" ")" "[" "]" "|" "/" "\"字符，都将进行转义处理，再进行匹配。此时，用转义后的路径进行激活态匹配才不会出现错误。

7.2.3　NavLink 源码解析

在实现上，NavLink 在基于 Link 实现正常的导航跳转的同时，增添了激活态判断的逻辑：

```
const NavLink = forwardRef(
  (
    {
      "aria-current": ariaCurrent = "page",
      activeClassName = "active",
      activeStyle,
      className: classNameProp,
      exact,
      isActive: isActiveProp,
      location: locationProp,
      strict,
      style: styleProp,
      to,
      innerRef,
      ...rest
    },
    // React v16.3 中的 forwardRef
    forwardedRef
  ) => {
    return (
      <RouterContext.Consumer>
        {context => {
          // 从传入的 locationProp 或 context 中获取 location
          const currentLocation = locationProp || context.location;
          // 创建 location 对象，其中有 pathname、search 等
          const toLocation = normalizeToLocation(
            // 兼容 to 为函数的情形
            resolveToLocation(to, currentLocation),
            currentLocation
          );
          const { pathname: path } = toLocation;
          // 转义特殊字符，如左右圆括号，如果不转义，则将无法匹配
          const escapedPath =
            path && path.replace(/([.+*?=^!:${}()[\]|/\\])/g, "\\$1");
          // 将 escaped 后的 path 进行匹配，注意没有大小写之分，所以 NavLink 的激
          // 活忽略大小写
          const match = escapedPath
```

```
      ? matchPath(currentLocation.pathname, {
         path: escapedPath,
         exact,
         strict
        })
      : null;
      // 使用 isActiveProp 函数或 match
    const isActive = !!(isActiveProp
      ? isActiveProp(match, currentLocation)
      : match);
      //计算 CSS 类
    const className = isActive
      ? joinClassnames(classNameProp, activeClassName)
      : classNameProp;
      // 计算 CSS 内联样式
    const style = isActive ? { ...styleProp, ...activeStyle } : styleProp;
    //  设置 Link 的 props
    const props = {
      // 无障碍属性，使得屏幕阅读器可识别
      "aria-current": (isActive && ariaCurrent) || null,
      className,
      style,
      to: toLocation,
      ...rest
    };
    // React v16.3 以上版本使用 forwardedRef
    if (forwardRefShim !== forwardRef) {
      props.ref = forwardedRef || innerRef;
    } else {
      // 在 React v15 中可使用 innerRef
      props.innerRef = innerRef;
    }
    // NavLink 使用 Link 进行导航
    return <Link {...props} />;
   }}
  </RouterContext.Consumer>
);
```

```
    }
  );
export default NavLink;
```

NavLink 在 Link 的基础上对传入的 to 属性进行了匹配操作。通过调用 matchPath 方法判断 to 是否与 currentLocation 的 pathname 匹配,如果匹配,则可以应用激活样式和 CSS 激活类等。这里需要注意,如果 NavLink 的 to 为 /a/b/c?search=1#somethin 等带 search、hash 的形式,内部也将转换路径得到 location,将仅使用 pathname 部分/a/b/c 与 currentLocation.pathname 进行匹配。注意,有一段 ref 的代码:

```
// 在 React v16.3 及以上版本中使用 forwardedRef,即外部属性上的 Ref 或 innerRef
if (forwardRefShim !== forwardRef) {
  props.ref = forwardedRef || innerRef;
} else {
  // 在 React v15 中可使用 innerRef
  props.innerRef = innerRef;
}
```

在 React v16.3 以上版本中开发可使用 Ref 属性。由于组件使用 forwardRef 进行包装,可将 Ref 直接传入内部的组件。同样,Link 也使用了 forwardRef:

```
// 在 React v16.3 中使用外部属性上的 Ref 或 innerRef
if (forwardRefShim !== forwardRef) {
  props.ref = forwardedRef || innerRef;
} else {
  props.ref = innerRef;
}
return <a {...props} />;
```

这样在 React v16.3 及以上版本中,无论是对于 Link 还是 NavLink,使用 Ref 或 innerRef 都可获得 a 标签对应的原生 DOM 节点引用。在 React v16.3 以下的版本中,仅能使用 innerRef 获得原生 DOM 节点引用。

7.3 DeepLinking

DeepLinking 仅在 React Native 中使用,其主要对 React Native 的 Linking 做一层封装。React Native 提供的 Linking 库通常用于调起其他 App 或本机应用。在 react-router-native 中,DeepLinking

是一个纯逻辑的组件，其不渲染任何 UI。DeepLinking 的主要作用为监听 Linking 的 URL 事件，并在事件响应函数中进行一次 push 操作：

```
import { Linking } from "react-native";
 Linking.addEventListener("url", this.handleChange);
handleChange = e => {
   this.push(e.url);
};
```

DeepLinking 通常在顶层树中实例化，监听 URL 的改变：

```
<NativeRouter>
    <DeepLinking />
 {/*…………*/}
</NativeRouter>
```

7.4　BackButton

同 DeepLinking 一样，BackButton 也仅在 React Native 应用中使用，其不渲染 UI，仅提供逻辑支持。其对 react-native 中的 BackHandler 进行一次封装：

```
// react-native 中的 BackHandler
import { BackHandler } from "react-native";
componentDidMount() {
    BackHandler.addEventListener("hardwareBackPress", this.handleBack);
}
 handleBack = () => {
   if (this.history.index === 0) {
     return false;
   } else {
     this.history.goBack();
     return true;
   }
 };
```

BackButton 监听 BackHandler 后退组件的 hardwareBackPress 事件，在事件响应函数中进行一次后退操作，以更新栈记录指针位置。

7.5 导航实战案例

7.5.1 为导航组件扩展路由匹配

传统的 Navlink 只能设计自身的激活样式，如 activeStyle、activeClassName 仅能设置 NavLink 中 a 标签的激活样式。但是，如果希望 NavLink 能将激活状态传递给子组件，传统的 NavLink 就无法做到了，这需要扩展 NavLink，如：

```
// NavLinkExt 将注入激活与否属性给 MenuItem
<NavLinkExt to="/user">
  <MenuItem />
</NavLinkExt>;
```

这时，MenuItem 可从 props.match 中获得是否匹配命中/user 路由，并可在 MenuItem 组件内根据路由匹配与否执行自身的渲染逻辑。

由于 Route 拥有匹配路由的能力，因此可以利用这一点，将 Route 与 NavLink 相结合：

```
import { createLocation } from "history";
import { __RouterContext as RouterContext } from "react-router";
import React from "react";
export const normalizeToLocation = (to, currentLocation) => {
  return typeof to === "string"
    ? createLocation(to, null, null, currentLocation)
    : to;
};
export function NavLinkExt(props: NavLinkProps) {
  const currentLocation = React.useContext(RouterContext).location;
  // 统一为 location 对象
  const toLocation = normalizeToLocation(props.to, currentLocation);
  return (
    /* 将 props 透传给 NavLink */
    <NavLink {...props}>
      <Route
        /* 同理，也将 exact、strict 等属性透传给 Route*/
        {...props}
        // path 为 toLocation 的 pathname 部分
        path={toLocation.pathname}
```

```
      // 无条件渲染子组件,并使用 cloneElement 将其注入 routeProps
      // 组件注入 routeProps 后,可在组件中读取 props.match
      children={(routeProps: RouteChildrenProps) => {
        return React.cloneElement(
          React.Children.only(props.children) as React.ReactElement,
          routeProps
        );
      }}
    />
  </NavLink>
);
}
```

将 NavLink 的 to 的 pathname 传递给 Route 的 path,这样可通过 Route 获得该 to 路径是否匹配命中;并使用 cloneElement 为 NavLink 的 children 注入 Route 的匹配 props,使用 Route 的 children,以保证子组件无条件渲染。这里使用 React.Children.only 保证了 NavLink 的 children 仅有一个,且也将 Navlink 的 props 透传给 Route,保证如 strict、exact 等属性也传递给 Route。

7.5.2 相对上下文路径导航组件

在 6.7.2 节曾介绍过相对路径 Route,使用相对路径 Route 可声明相对父级 Route 的 path。同样,在一些开发场景中,使用 Link 或 NavLink 的导航,也有不希望关心整个导航路径的情况,而仅需关心相对于上下文路径的相对路径。如:

```
// 若上下文匹配路径为/a,则单击此导航将导航到/a/c
<RelativeLink to={`/c`}>相对导航</RelativeLink>
```

相对于当前匹配路径,RelativeLink 以相对路径进行导航,如当前匹配路径为/a,则 RelativeLink 可导航到/a/b、/a/c 等。实现如下:

```
import { __RouterContext } from "react-router";
import { Link, LinkProps } from "react-router-dom";
import * as React from "react";
import { useContext } from "react";
export default function(props: LinkProps) {
  // 获取上下文中的命中信息
  const { match } = useContext(__RouterContext);
  if (match) {
    // 拼接匹配路径与相对导航路径
```

```
    return <Link {...props} to={`${match.url}${props.to}`} />;
  } else {
    return <Link {...props} />;
  }
}
```

使用 useContext(__RouterContext)获得组件树中最近的 Route 内 Provider 所提供的 Context 值。如果上级 Route 匹配命中，match 不为 null，则可使用上级 Route 的真实匹配结果 match.url，再拼接 props.to，得到最终的 Link 导航路径。如果上级 Route 匹配未命中，match 为 null，则此时可返回原 Link 组件，不做任何逻辑处理，保留了导航功能，此时的组件等同于原 Link 组件。

7.5.3 相对上下文路径的导航方法

对于原始的 history.push、history.replace 方法，在进行跳转时需要提供完全路径。在一些层级较深的组件内，如果希望相对上下文路径进行跳转，那么为 history 方法提供完全路径则会显得臃肿，如果能像操作目录一样操作导航跳转，如：

```
// 相对所在 Route 返回上级路径进行/list 导航
history.push({pathname: '../list'}, true)
// 相对所在 Route 所在路径进行/detail 导航
history.push('/detail',state,true)
// 相对所在 Route 返回上级路径进行/list 导航
history.replace('../list', state, true)
// 相对所在 Route 所在路径进行/detail 导航
history.replace({pathname: '/detail'}, true)
```

最后一个参数为 true 表明导航的路径是相对路径，相对的基准为所在 Route 的命中 path。引入相对路径的 history 导航将更加简洁且符合操作习惯。最后一个参数不传入时默认为 undefined，这时导航也将与 history 的导航行为一致。

原始 history 支持相对当前地址的跳转，如果为 history 方法提供相对路径，则 history 将以当前 history.location 为基准进行相对导航，相关内容可查看 1.2.3 节。

为实现相对上下文路径导航，可在 Route 使用 render 或 children 渲染组件时修饰原 history，返回新 history 后注入组件中：

```
<Route
    path={'/user'}
    render={routeProps => {
```

```
        // 修饰 history
        const newHistory = decorateHistory(routeProps.history, routeProps.match.url)
        return <User {...routeProps} history={newHistory}/>
      }}
    />
```

上例在 render 中，通过调用 decorateHistory 将原 history 对象传入，并将当前 Route 匹配命中的 URL 传入，作为基准导航部分。decorateHistory 返回得到的新 history 对象的属性将与原有 history 的属性一致，之后将新返回的 history 传入组件中，在组件中使用 history 即可获得新的相对上下文路径导航的能力。

decorateHistory 方法将保存原始 push 与 replace 方法，并基于原始方法构造新的可相对导航的 push 和 replace 方法，且构造了新的 history 对象返回：

```
function decorateHistory(history: History, currentRoutePath: string) {
  const methods = ["push", "replace"];
  const originHistoryMap = {};
  methods.forEach(method => {
    // 保存原 history 方法
    originHistoryMap[method] = history[method];
  });
  const newHistory = {};
  methods.forEach(method => {
    newHistory[method] = function(...args) {
      // 合并路径
      const { path, state } = concatLocation(args, history, currentRoutePath);
      // 调用 resolve 方法处理 ".." "." 情况，并调用原 history 方法
      return originHistoryMap[method](resolve(path), state);
    };
  });
  // 返回新的 history 对象
  return {
    ...history,
    ...newHistory
  };
}
```

上例需要返回新的 history 对象，而不能在原 history 对象上修改。原因是此时对 history 对象

的修饰不具有通用性，因为每个传入组件的 history 对象的相对导航基准即 Route 的 path 是不一样的。

上例在新的 push、replace 方法中使用了 concatLocation 函数。concatLocation 函数用于判断 args 中最后一个参数 shouldUseRelativePath，如果为 true，则拼接 concatLocation 函数传入的基准路径与参数 args 中的相对路径，并进行返回：

```
function concatLocation(args: any[], history:History, currentRoutePath:string) {
  let realpath:string;
  let state:any;
  // 在 args 的第一个参数，即 location 为对象的情况下进入此分支
  if (typeof args[0] === "object") {
    const location = args[0];
    const shouldUseRelativePath:boolean = args[1];
    realpath = history.createHref(location);
    state = location.state;
    if (shouldUseRelativePath) {
      // 拼接基准路径与相对路径
      realpath = history.createHref({
        ...location,
        // currentRoutePath 为基准路径
        pathname: `${processPath(currentRoutePath)}/${processPath(
          location.pathname
        )}`
      });
    }
  } else {
    // location 为字符串的情况
    realpath = args[0];
    state = args[1];
    const shouldUseRelativePath:boolean = args[2];
    if (shouldUseRelativePath) {
      // 拼接基准路径与相对路径
      realpath = `${processPath(currentRoutePath)}/${processPath(realpath)}`;
    }
  }
  return { path: realpath, state };
}
```

如果拼接后的相对路径为/a/b/c/../d，则需要把路径变换为/a/b/d，这可通过以"/"分隔数组来实现：

```
export function resolve(path: string) {
  if (path.includes("..") || path.includes(".")) {
    const segments = path.split("/");
    let resolvedSegments:string[] = [];
    segments.forEach((segment: string) => {
      if (segment === "..") {
        // ".."表示上一级路径，执行出栈操作
        resolvedSegments.pop();
      } else if (segment !== ".") {
        resolvedSegments.push(segment);
      }
      // 如果为"."，则什么都不做，继续进行下一个段落
    });
    return resolvedSegments.join("/");
  } else {
    return path;
  }
}
```

7.5.4　为导航组件扩展 search 和 hash 支持

在 7.1.2 节 Link 源码解析中，曾介绍了 Link 组件底层使用了 history 的导航方法，因而对于 history 的方法及参数类型，在 Link 组件中都适用。

同时在 2.2.2 节中，曾介绍过导航方法允许传入参数路径：

```
history.push('?bar'); // 本级路径不变，设置 query
history.push('#foo'); // 本级路径不变，设置 hash
```

因此，对于 Link 组件，天生支持 to 参数改变 search 和 hash：

```
<Link to="?c=1">改变 search</Link>
<Link to="#page1">改变 hash</Link>
```

但原始的 history 在改变 search 和 hash 时存在一个问题，即 history 不能分辨 search 与 hash 的区别，其在设置路径时，仅相对当前路径进行设置，这会清除某些本来生效的 search 或 hash：

```
// 若当前地址为/user#somehash，如下导航，地址将变为 /user?c=1，#somehash 被清除
```

```
<Link to="?c=1">改变 search</Link>
// 若当前地址为/user?c=1,如下导航,地址将变为 /user#page1, ?c=1 被清除
<Link to="#page1">仅改变 hash</Link>
```

为了解决这个问题,需要为 Link 提供支持,使得 Link 能识别 search 与 hash,并仅对需要的部分进行替换。扩展后的 Link 将支持仅对 search 或 hash 进行替换:

```
// 若当前地址为/user#somehash,如下导航,地址将变为 /user?c=1#somehash
<Link to="?c=1">仅改变 search</Link>
// 若当前地址为/user?c=1,如下导航,地址将变为 /user?c=1#page1
<Link to="#page1">仅改变 hash</Link>
```

在实现上,如果 Link 的 to 参数的字符以 "?" 或 "#" 开头,则可知 Link 的意图为改变 search 或 hash。原始 Link 组件不支持这一特性,但由 7.1.2 节源码可知,Link 可执行 onClick,并可以阻止默认导航行为。利用这一特性,可以封装一个支持改变 search 与 hash 的扩展 Link。注意,扩展的 Link 在浏览器路由与 hash 路由中同样适用。

```
import * as React from "react";
import { Link, LinkProps } from "react-router-dom";
import { __RouterContext } from "react-router";
import { History } from "history";
import * as queryString from "query-string";
// 引入 2.7 节的 search、hash 替换方法
import { replaceSearch, replaceHash } from "./util/history";
// 获取当前 search
function useSearch() {
  const location = React.useContext(__RouterContext).location;
  const { search } = location;
  return React.useMemo(() => queryString.parse(search), [search]);
}
interface LinkPropsExt extends LinkProps {
  // 是否不关心当前地址中的 search,完全替换
  replaceSearch?: boolean;
}
export default function LinkExt(props: LinkPropsExt) {
  // 获取 history,可使用其导航方法
  const history = React.useContext(__RouterContext).history;
  // 获取当前 search 备用
  const currentSearch = useSearch();
```

```
const onClick = React.useCallback(
  (event: React.MouseEvent<HTMLAnchorElement>) => {
    props.onClick && props.onClick(event);
    if (typeof props.to === "string") {
      // 路径以 search 的形式以 "?" 开头
      if (props.to.startsWith("?")) {
        /*
          如果路径以 "?" 开头，且包括 "#"，如传入的路径为 "?a=1#hash"，则可不做处理，
          Link 的原始逻辑能满足同时改变的需求
        */
        if (!props.to.includes("#")) {
          event.preventDefault();
          if (props.replaceSearch) {
            // 全量覆盖，不关心当前 search 的情况，replaceSearch 可查看 2.7 节
            replaceSearch(history, props.to);
          } else {
            const toSearch = queryString.parse(props.to);
            // 在当前 search 的基础上增量覆盖
            const search = { ...currentSearch, ...toSearch };
            replaceSearch(history, search);
          }
        }
      }
      if (props.to.startsWith("#")) {
        event.preventDefault();
        // 仅覆盖 hash, replaceHash 可查看 2.7 节
        replaceHash(history, props.to);
      }
    }
    return;
  },
  [props.to, props.replaceSearch, currentSearch, history]
);
// 透传属性到 Link
return <Link {...props} onClick={onClick} />;
}
```

上例自定义了 Hooks：useSearch，用于获取当前 history 的 search 值。这里使用路由上下文获

取当前的 location 对象，再从对象中获取到 search 字符串，只关心路由上下文的此种方式能兼容浏览器路由和 hash 路由多种场景。search 的获取由 history 提供。由于返回的 search 有可能变动频率较低，这里可以使用 React.useMemo 进行记录，useMemo 的依赖为 location 中的 search。当 search 不变时，useMemo 返回之前计算过的值；当 search 变化时，useMemo 重新计算并进行返回。

扩展 Link 的核心在于传入自定义的 onClick。在 onClick 中，判断 props.to 的前缀，如果为"?"或"#"开头，则执行改变 search 或 hash 的逻辑，并阻止原生 Link 的导航行为；而如果路径同时包含"?"与"#"，则可不做处理，Link 的原始逻辑能满足同时替换两者的需求。

与此同时，为了灵活地控制 search，可设置一个 replaceSearch 属性用以判断 search 的改变方式，如增量覆盖或全量覆盖。在参数的替换执行上，最终的替换由 history 执行。执行 history.replace 方法，只允许传入 search 和 hash，保证 pathname 不受影响，在 2.7 节曾有过介绍。并且由于从路由上下文中获得单例的 history 对象，参数替换由 history 负责，由 history 进行 search 或 hash 的改变，这同时适用于浏览器路由与 hash 路由等多种场景。注意，onClick 使用了 React.useCallback 方法，目的是在依赖不变时保存 onClick，不用每次渲染都重新生成一份 onClick 实例，仅在 props.to、props.replaceSearch、currentSearch、history 中的任意一个依赖有变化时，才重新生成 onClick 函数。

7.6 小结

本章主要介绍了 React Router 中导航组件的使用，包括 Link 基本导航组件、可判断激活态的 NavLink 组件，以及 React Native 中的 DeepLinking 与 BackButton 组件。导航组件作为导航命令的包装组件，并不真正具备导航能力，它们的导航能力都依赖上下文中的 history。第 2 章曾介绍过 history 的使用及其原理，本章在组件基本使用的基础上，通过源码的解析介绍，帮助读者加深对 history 的理解。与此同时，本章通过实战案例的引入，也帮助读者学习、巩固导航组件的使用方法，为理解和掌握 React Router 打下良好的基础。

参考文献

https://github.com/ReactTraining/react-router.

第 8 章
其他路由组件及方法

在第 5、6、7 章中，分别介绍了 React Router 应用中不可缺少的三大元素，包括 Router、Route 及 Link。合理使用以上三大元素即可满足 React 应用的正常运行。但在某些场景下，恰当使用某些辅助组件或方法，将更能提升开发效率和代码质量。React Router 为此提供了多个组件及方法，如路由匹配组件 Switch、mathPath 路径匹配方法等。本章将介绍 React Router 中的各辅助组件及方法，以帮助读者更便捷、高效地开发 React 应用，同时完善对 React Router 整个体系的介绍。

8.1 Switch

8.1.1 Switch 简介

Switch 组件是一种路由匹配组件，使用 Switch 需要搭配 Route 组件。Switch 组件接收的 props 如下：

```
export interface SwitchProps {
    children?: React.ReactNode;
    location?: H.Location;
}
```

在 Switch 接收的 SwitchProps 中，仅接收 children 子节点与 location 地址。Switch 在配置上不复杂，本节将介绍 Switch 的功能与基本使用方法。

Switch 拥有挑选 Route 的能力，会挑选并渲染第一个匹配路由路径的 Route。当某 Route 匹配命中时，其余未匹配命中或者即便匹配路径的 Route，都会返回 null，Switch 只渲染第一个匹配命中的 Route，如：

```
<Switch>
    <Route path="/([ab])" component={A} />
    <Route path="/a" component={B} />
    <Route path="/b" component={C} />
</Switch>
```

当路由路径为/a 或/b 时，仅有 A 组件能渲染，Switch 保证了第一个匹配成功的 Route 得到渲染，其他 Route（即便也能匹配成功）将被忽略。从功能上说，Switch 的能力类似 React.children.only，都只保证渲染一个子组件。

如果 Switch 组件的任一子组件 Route 都没有命中路由，且由上级传入的路由上下文中也没有匹配命中的 Route，则 Switch 组件中的 render 方法将返回 null，不渲染子组件。

在开发者使用 Switch 组件时，仅有小于或等于一个的子组件 Route 或 Redirect（将在 8.2 节介绍）能得以渲染。对于越放在靠前位置的 Route，其有越高的渲染优先级。

因此，对于 Switch，较为普遍的用法是把匹配范围较广的 Route 放在 Switch 子组件的最后位置，渲染如 "404" 等通用页面。如匹配范围最广的 "/" 路径，将对所有路径匹配（在没有 exact 属性的情况下），则将路径为 "/" 的 Route 放在子组件的最后位置，Switch 将会优先渲染在 "/" 之前命中的 Route 组件，仅在除 "/" 外的 Route 都匹配失败时才会命中路径为 "/" 的 Route，渲染对应的通用组件：

```
// 优先渲染/a，当其未命中时才渲染/
<Switch>
    <Route path="/a" component={A} />
    <Route path="/" component={Page404} />
</Switch>
```

在 Switch 的使用上，需要结合具体场景。若没有 Switch，如：

```
// 在没有 Switch 的场景中，在/a 路径下，A 与 B 组件同时得到渲染
<>
    <Route path="/([ab])" component={A} />
    <Route path="/a" component={B} />
</>
```

此刻若有路径/a，则两个 Route 都匹配成功，A 与 B 组件都将得到渲染展示，这在某些场景中也是需要的。

8.1.2 Switch 源码解析

Switch 的源码很短，如下所示：

```
class Switch extends React.Component {
  render() {
    return (
      <RouterContext.Consumer>
        {context => {
          // Switch 可从 props 中接收 location,这在动画等场景中会用到
          const location = this.props.location || context.location;
          // 用以记录匹配成功的子组件
          let element, match;
          React.Children.forEach(this.props.children, child => {
            if (match == null && React.isValidElement(child)) {
              element = child;
              // 从 Route 中读取 path 或者从 Redirect 中读取 from
              const path = child.props.path || child.props.from;
              // 当 match 不为 null,即第一次匹配到时,即会记录 match
              match = path
                // 使用 path-to-regexp 的路径匹配
                ? matchPath(location.pathname, { ...child.props, path })
                : context.match;
            }
          });
          // 保存原子组件的属性,并将 Switch 的 location 及计算出的 computedMatch
          // 传入子组件
          return match
            ? React.cloneElement(element, { location, computedMatch: match })
            : null;
        }}
      </RouterContext.Consumer>
    );
  }
}
```

在 Switch 实现上，遍历其每一个 child 子组件，获取 child 子组件 props 上的 path 属性后（对 Redirect 组件将获取 from 属性），再跟上下文中的 location 进行匹配对比。若匹配成功，则使用

React.cloneElement 方法克隆原组件，以保持原 child 子组件的 props，并传入当前的 location 和已经计算出来的结果变量 computedMatch。

为什么源码中不使用 React 的 React.Children.toArray 方法转换 children，而直接使用 forEach？这里要考虑同一个组件渲染在不同 URL 中的情况，如下所示：

```
<Switch>
  <Route path="/a" component={A} />
  <Route path="/b" component={A} />
</Switch>
```

在路径/a、/b 同时渲染同一个 A 组件的情况下，若当前的路径为/a，并从该路径导航到/b 路径，原先/a 路径命中并渲染过 A 组件，且导航到/b 路径也同样渲染 A 组件。由于对 Switch 的子组件来说，将同样渲染 Route，Route 也没有 key 的变化，Route 的渲染也没有发生变化（都渲染 A 组件），因此 A 组件并不会触发 componentWillUnmount，而是会进入 A 组件更新的生命周期。如果源码使用 React.Children.toArray 方法，由于该方法会为组件增加 key 标志，所以这时 Route 会因为 key 的不同，使旧 key 对应的 Route 被销毁，新 key 对应的 Route 被挂载。这样的销毁和挂载过程会导致同一个 A 组件也被销毁与重新挂载。

如果希望每次命中路由都能销毁旧组件，并重新渲染进而执行 componentDidMount 生命周期方法，则可以为渲染相同组件的各 Route 加入唯一的 key 值，如下所示：

```
// 当导航时，A 组件会先销毁，再重新渲染并执行 componentDidMount 生命周期方法
<Switch>
  <Route path="/a" key="a" component={A} />
  <Route path="/b" key="b" component={A} />
</Switch>
```

这时，由于 key 值不同，当导航从/a 到/b 时，key 值为 b 对应的 Route 将得到渲染，但是由于原先的 Route 的 key 值为 a，key 值不一致，所以按照 React 的 diff 机制，key 值为 a 对应的 Route 将会被销毁，key 值为 b 对应的 Route 将会被挂载。对应的 A 组件也会执行相同的操作，即 A 组件会被销毁，并重新渲染，会执行 componentDidMount 生命周期方法。

若在上例中没有 Switch 组件包裹 Route，如：

```
// 当导航时，A 组件会执行 componentWillUnmount 生命周期方法
<>
  <Route path="/a" component={A} />
  <Route path="/b" component={A} />
</>
```

由于没有父组件 Switch 的记录逻辑，每次导航 Route 未命中时对应组件都会被销毁。当导航从路径/a 到/b 时，离开的 A 组件会执行 componentWillUnmount 生命周期方法，进入路径的 A 组件会执行 componentDidMount 生命周期方法。

注意，Switch 会影响其子组件 Route 的 location。对于 Route，如果其被 Switch 组件包裹，则 Route 的 location 将为 Switch 组件中的 location。Switch 组件中的 location 有 2 个来源，一个是从 props.location 中获取，另一个是从 context.location 中获取。如果为 Switch 设置了 location，则需要注意这里设置的 location 也将影响到子组件的 location。如：

```
const location = {
  pathname: "/foo/baz",
  search: "",
  hash: "",
  state: {}
};
<Switch location={location}>
  {/* Route 将使用传入 location 的 pathname，Route 将匹配成功 */}
  <Route path="/foo/baz" component={FooBaz} />
</Switch>;
```

8.2 Redirect

Redirect 组件在 React Router 中为重定向组件，其提供了组件级别的导航行为。作为 React Router 中的常用组件，Redirect 组件能提供基本跳转与条件跳转两类导航行为。

8.2.1 基本跳转

Redirect 组件接收如下 props：

```
export interface RedirectProps {
    // 重定向导航地址
    to: H.LocationDescriptor;
    // 是否以入栈方式导航
    push?: boolean;
    // 条件跳转可用属性，用于指明从哪个地址触发重定向，需要搭配 Switch 组件
    from?: string;
```

```
  // 等同于 from 属性，一般不使用
  path?: string;
  // 在使用 from 且搭配 Switch 组件时，用于指明触发重定向时 from 需要绝对匹配
  exact?: boolean;
  // 在使用 from 且搭配 Switch 组件时，用于指明触发重定向时 from 需要严格匹配
  strict?: boolean;
  // 在使用 from 且搭配 Switch 组件时，用于指明触发重定向时 from 对大小写敏感
  sensitive?: boolean;
}
```

同 Link 类似，Redirect 的基本使用仅需要传入 to 属性标识要导航到的路径：

```
<Redirect push to="/user/profile" />
```

to 除了可以为字符串，还可以为地址描述对象：

```
<Redirect
  to={{
    pathname: "/user",
    search: "?c=1",
    state: { a: 1 }
  }}
/>
```

对于 Redirect 组件，需要关注其生命周期。当 Redirect 组件挂载触发 componentDidMount 生命周期时，会执行重定向操作。当 Redirect 组件进入更新的生命周期时，若新 props 的 to 属性与原 props 的 to 属性不相等，则同样会执行重定向操作。

Redirect 默认的导航方式是替换历史栈的导航方式，如果传入 props 中 push 为 true，则将会执行 push 方法，会入栈一个路由地址。

8.2.2 条件跳转

Redirect 的基本跳转方法并没有能命中一个路由路径再跳转的能力，比如希望能在命中路径 /a 的同时再跳转到路径/b。为了使得 Redirect 支持条件跳转，可以使用 RedirectProps 中的 from 属性，作为 Redirect 组件在条件跳转时的触发地址。但要注意一点，from 属性需要配合 Switch 组件来使用，如：

```
<Switch>
  <Redirect from='/old-path' to='/new-path'/>
```

```
    <Route path='/new-path' component={Place}/>
</Switch>
```

上例在当前路径为/old-path 时，Redirect 会将地址导航到/new-path 路径。

对于有参数跳转的 Redirect，如：

```
<Switch>
  <Redirect from='/old-path/path/:id' to='/new-path/:id'/>
  <Route path='/new-path' component={Place}/>
</Switch>
```

经由 Switch 计算的 from 的参数，会填充到 to 中进行跳转，如原路径为/old-path/path/1，则得到参数 id 为 1，会填充到 to 的 id 参数中，因而会跳转到/new-path/1 这个路径中。

在 Redirect 父组件为 Switch 的条件跳转场景中，可以设置 exact、strict、sensitive 等属性。因为涉及父级 Switch 对子组件 Redirect 的 from 属性做路径匹配，所以会使用 Redirect 的 exact、strict、sensitive 属性参与到路由匹配的计算中。带 from 属性的 Redirect 与 Route 的行为一致，因而带 from 属性的 Redirect 可以当作特殊的 Route，from 属性在性质上就是 Route 的 path 属性。

要注意，父组件为 Switch 的 Redirect 如果没有 from 属性，就会跟无 path 的 Route 一样，使用上下文中的匹配结果。如果上下文中的 Route 匹配成功，那么这里就会渲染，如：

```
<Router history={browserHistory}>
  <Switch>
    <Redirect to="/a" />
    <Route path="/a" component={A} />
  </Switch>
</Router>
```

Redirect 会匹配成功并且渲染（Router 提供的上下文默认匹配成功），将应用导航到/a 路径上。

在上例中，由于 Switch 只渲染第一个匹配的组件，这里为第一个 Redirect 组件，因而第二个 Route 组件无法得到渲染，A 组件永远无法展示，这样页面地址会停留在/a 路径，但无具体 A 组件内容。

8.2.3 源码解析

在 Redirect 的源码实现中，通过路由上下文获取到 history 对象与 staticContext 静态上下文。通过获取到的 history 对象，可以使用 push 或 replace 方法完成重定向的导航工作。获取

staticContext 的意图是判断导航应该触发的时机，如有 staticContext，则说明此时的 Redirect 组件被用于服务端。由于服务端不会触发挂载的生命周期，因此若希望在服务端完成重定向，则导航需要在 render 中进行。Redirect 源码分析如下：

```
function Redirect({ computedMatch, to, push = false }) {
  return (
    <RouterContext.Consumer>
      {context => {
        const { history, staticContext } = context;
        // Redirect 的导航模式
        const method = push ? history.push : history.replace;
        const location = createLocation(
          computedMatch
            ? typeof to === "string"
            // 结合 Switch 组件使用，父级 Switch 组件计算出 computedMatch
              // 将 from 部分的命名参数传递到 to
              ? generatePath(to, computedMatch.params)
              : {
                  ...to,
                  pathname: generatePath(to.pathname, computedMatch.params)
                }
              // 不能单独使用/user/:id，需结合 Switch 组件
            : to
        );
        //staticContext 即静态 Router 场景，在 5.6 节曾介绍过静态 Router
        if (staticContext) {
          method(location);
          return null;
        }
        return (
          <Lifecycle
            onMount={() => {
              // 初始挂载成功，进行一次导航
              method(location);
            }}
            onUpdate={(self, prevProps) => {
              const prevLocation = createLocation(prevProps.to);
              if (
```

```
            // 在 update 阶段，不相等，重新导航
            !locationsAreEqual(prevLocation, {
              ...location,
              key: prevLocation.key
            })
          ) {
            method(location);
          }
        }}
        to={to}
      />
    );
  }}
</RouterContext.Consumer>
);
}
export default Redirect;
```

从源码中可看到，若 Redirect 组件结合了 Switch 组件，则可使用参数化重定向。在创建重定向 location 时，会使用父级 Switch 组件根据 from 属性计算出的 computedMatch，其中包含了 from 属性的命名参数。

在重定向时机上，Redirect 组件在挂载成功后会无条件地使用 to 属性进行导航跳转。而当 to 有变更时，同样也会使用最新的 to 进行路由跳转（源码中 Lifecycle 仅是 React 生命周期的简单包装）。

结合 8.1.2 节 Switch 的源码可知，from 属性需要结合 Switch 组件来使用。在 Switch 渲染子组件 Redirect 时，会获取 Redirect 的 from 属性值。若 from 属性值匹配命中路径，则可渲染 Redirect 组件。事实上，当 Redirect 结合 Switch 使用时，可以在 Redirect 中使用 path 属性，效果与 from 属性一致，这使得 Redirect 在表现上更像一个 Route；但不建议这样使用，因为 from 属性相对于 path 属性更具备条件跳转的语义。

当 Redirect 不结合 Switch 使用时，from 属性将不生效，但可以在业务中根据当前地址信息判断 Redirect 组件是否应当渲染。

如果希望单独使用 Redirect（不结合 Switch）时可以使用 from 属性，则可以封装一个组件，提供条件跳转：

```
// 当前路径为/a 时再进行跳转
<ConditionalRedirect from='/a' to="/b">
```

```
// 无条件跳转到路径/b
<ConditionalRedirect to="/b">
```

其实现如下：

```
function ConditionalRedirect(props){
  // 提取出 location 供 Switch 使用，如某些需要缓存 location 的场景
  const {location,...rest} = props;
  if(props.from){
    return <Switch location={location}>
      // from 与 to 都传入 Redirect 中
      <Redirect {...rest}/>
    </Switch>
  }else{
    return <Redirect {...rest}/>
  }
}
```

上例实现的 Redirect 在组件内部使用了 Switch，屏蔽了 Switch 的路径匹配逻辑。

此外，还能设计诸如强制更新的 Redirect：

```
function ForceRedirect(props){
  const key = Math.random();
  return <Redirect key={key} {...props}/>
}
```

利用每次渲染设置不同的 key，使得 Redirect 组件每次渲染都被销毁并再次挂载，从而进行跳转。

8.3　Prompt

Prompt 组件称为跳转确认组件，用于在用户导航时提示用户是否确认导航到某路径。Prompt 组件基于 history.block 方法（在 2.1 节曾介绍过其用法），其用法如下：

```
import { Prompt } from "react-router";
<Prompt when={true} message="是否导航到/a" />
```

Prompt 组件的 props 类型如下：

```
export interface PromptProps {
  message: string | ((location: H.Location) => string | boolean);
```

```
  // 当 when 为 true 时,Prompt 才有跳转确认行为
  when?: boolean;
}
```

当 when 为 true 时,如果没有设置确认框,则会弹出系统默认的确认框。系统默认的确认框中的信息为 Prompt 组件的 message 字段的内容:

```
<Prompt when={true} message="确认是否离开页面?" />
```

当 message 为函数时,函数参数中 location 为即将跳转到的 location 地址,如果函数返回 true,则允许跳转;如果函数返回字符串,则会弹出确认框提示用户进行操作(when 需要为 true),如:

```
<Prompt
  when={true}
  message={location =>
    location.pathname.startsWith("/app")
      ? true
      : `确认是否前往 ${location.pathname}?`
  }
/>
```

注意,如果在创建 history 时传入了 getUserConfirmation:

```
const history = createBrowserHistory({
  getUserConfirmation: (messge, callback) => {
    /** ............ 自定义弹窗 */
  }
});
```

在应用的 Router 使用了这个 history 的情况下,Prompt 组件将使用此自定义弹窗。

在 Prompt 源码实现上,Prompt 使用 history 的 block 函数(2.5.5 节曾介绍过 history.block 运行原理)进行跳转确认:

```
function Prompt({ message, when = true }) {
  return (
    <RouterContext.Consumer>
      {/** 使用 RouterContext.Consumer 的目的是获取 history 对象 */}
      {context => {
        // 当 when 为 false 时,返回 null,Lifecycle 组件会被销毁,执行 onUnmount,
        // 取消 history.block 的注册
        // 服务端路由不用渲染 Prompt,导航确认不在服务端起作用
        if (!when || context.staticContext) return null;
```

```
            const method = context.history.block;
            return (
              <Lifecycle
                onMount={self => {
                  // method 为 history.block
                  self.release = method(message);
                }}
                onUpdate={(self, prevProps) => {
                  if (prevProps.message !== message) {
                    // message 不同，先释放原 block，再更新引用
                    self.release();
                    self.release = method(message);
                  }
                }}
                onUnmount={self => {
                  // release 保存的是 history.block 的返回结果，需要释放
                  self.release();
                }}
                message={message}
              />
            );
          }}
        </RouterContext.Consumer>
    );
}
export default Prompt;
```

从源码中可以看到，Prompt 仅在 when 属性为 true 且在非服务端场景中时才具有拦截效果。

同时，在 when 为 true 的前提下，如果渲染 Prompt 前后两次的消息不一致，则会更新弹窗的信息。在实现上，需要先释放原先的注册函数，并重新绑定。

8.4 withRouter

withRouter 是 React Router 提供的高阶组件。在一些处于很深层级的组件中，如果希望获得 props.history、props.location 等对象，又不希望从上层逐级传入，则可使用 withRouter 高阶组件注入相关 RouteComponentProps 属性：

```
export interface RouteComponentProps<Params extends { [K in keyof Params]?:
string } = {}, C extends StaticContext = StaticContext, S = H.LocationState> {
  history: H.History;
  location: H.Location<S>;
  match: match<Params>;
  staticContext?: C;
}
```

若现有 Info 组件位于组件树深处，则可通过调用 withRouter 获得新的组件：

```
export default withRouter(Info)
```

上例会为 Info 注入类型 RouteComponentProps 中的所有属性。

在 withRouter 源码中会使用到 RouterContext，由于 React Router 也将 Context 通过 __RouterContext 暴露，withRouter 也可自行实现，其源码较简单：

```
function withRouter(Component) {
  const displayName = `withRouter(${Component.displayName || Component.name})`;
  const C = props => {
    const { wrappedComponentRef, ...remainingProps } = props;
    return (
      <RouterContext.Consumer>
        {context => {
          return (
            <Component
              {...remainingProps}
              {...context}
              ref={wrappedComponentRef}
            />
          );
        }}
      </RouterContext.Consumer>
    );
  };
  C.displayName = displayName;
  C.WrappedComponent = Component;
  return hoistStatics(C, Component);
}
```

使用 RouterContext.Consumer 消费组件树中距离最近的一个 RouterContext.Provider 所提供的值，即组件树中最近的一个 Route 所提供的值。得到 Context 之后，再渲染传入的组件，为传入的组件注入 Context 上的属性。最后的返回值使用 hoistStatics 将原组件的静态方法也同样复制到 C 进行返回，这是高阶组件常需要注意的。源码中上下文的获取也可使用 React Hooks，改写较简单，这里不再特意实现，感兴趣的读者可自行完成。

8.5　matchPath

matchPath 为 path-to-regexp 的方法做了一层包装。其在源码内部大量使用，用于判断如 Route 的 path 路径是否与 URL 路径相匹配。

matchPath 的调用方式如下：

```
matchPath("/a/b/c", {
  path: "/a/b",
  exact: false,
  strict: false,
  sensitive: false
});
```

第一个参数一般为当前的 location 路径地址，第二个参数为匹配配置。

matchPath 方法所返回的结果即 props.match 对应的结果：

```
export interface match<Params extends { [K in keyof Params]?: string } = {}> {
  params: Params;
  isExact: boolean;
  path: string;
  url: string;
}
```

事实上，Route 内部的 props.match 就是靠此方法计算获得的。

在本节示例中，matchPath 的执行结果如下：

```
{ isExact: false, params: {}, path: "/a/b", url: "/a/b" }
```

在源码实现上，matchPath 为 path-to-regexp（在 6.2.1 节曾有过简单介绍）的方法做了一层包装，真正的路径匹配由 path-to-regexp 进行。

matchPath 配置（Route 的选项配置）到 path-to-regexp 选项配置的映射关系如表 8-1 所示。

表 8-1　matchPath 配置到 path-to-regexp 选项配置的映射关系

matchPath 配置	path-to-regexp 选项配置
exact	end
strict	strict
sensitive	sensitive

8.6　实战案例

8.6.1　路由动画

在需要为路由加入动画的场景中，可使用 react-transition-group 库。从本质上来说，react-transition-group 库不具备管理动画的能力，其仅为开发者提供了一系列用于动画过渡状态管理的组件，如 Transition 组件为开发者提供了可描述动画的状态机；CSSTransition 组件基于 Transition 组件更进一步为开发者提供了动画各阶段 CSS 类型；TransitionGroup 组件基于子组件 key，管理了多个 Transition 或 CSSTransition 子组件的动画过渡状态。对于路由动画，通常可使用 CSS 动画进行，可使用 CSS 类描述路由的进入和退出动画。具体到使用 react-transition-group 库上，描述动画一般有两种思路：第一种是使用 TransitionGroup 组件管理多个路由页面，其特点是不需要人工触发动画，仅需要为过渡组件（一般为 CSSTransition 组件）提供唯一的 key 值；第二种是使用如 CSSTransition 组件人工触发动画，这种实现方式由于没有 TransitionGroup 组件的管理，虽然不需要唯一的 key 值，但是需要人工控制组件动画的触发。在开发场景中，这两种实现方式都可实现路由动画，可在具体开发场景中选择其一。下面对这两种实现方式进行简要介绍。

如果使用 TransitionGroup 组件控制动画，则需要先了解 TransitionGroup 组件的作用原理，以便更好地实施路由动画。TransitionGroup 作为管理组件，其不提供管理动画的能力，也不为组件提供动画各阶段的回调。其本质是管理即将被销毁的组件与即将挂载的组件，通过管理子组件的各个 key 值实现。在 TransitionGroup 组件内部，TransitionGroup 会保存前后两次渲染的 children 元素，通过 key 的比对可获知哪些元素需要触发进入的动画，哪些元素需要触发退出的动画。以一个实例来说明，如 A 组件实例从 key 为 1 变成 key 为 2：

```
<TransitionGroup>
  <A key='1'/>
```

```
{/* 从某个时刻开始，从 key=1 变为 key=2 */}
{/* <A key='2'/> */ }
</TransitionGroup>
```

虽然上述过程仅有一个 A 组件参与，仅对 A 组件改变了 key 值，但父组件 TransitionGroup 通过 React.cloneElement 的方式克隆保留了两个时刻的元素信息，即即将被销毁的<A key="1"/>元素与即将挂载的<A key="2"/>元素。父组件可以同时渲染两个子组件，即<A key="1"/>与<A key="2"/>，并对<A key="1"/>加入退出动画，对<A key="2"/>加入进入动画。基于以上分析，可实现如下路由动画：

```
<TransitionGroup>
    <CSSTransition
      // 每次 location 更新，key 不同
      key={location.key}
      // 应用 CSS 动画
      classNames="fade"
      // 子组件保留时间
      timeout={300}
    >
      {/* 传入 location 形式，使得父组件能缓存 */}
      <Switch location={location}>
        <Route exact path="/a" component={A} />
        <Route exact path="/b" component={B} />
      </Switch>
    </CSSTransition>
</TransitionGroup>
```

在上例中，CSSTransition 组件将对进入渲染的 DOM 元素应用 fade-enter、fade-enter-active、fade-enter-done 的 CSS 类。其中，fade-enter、fade-enter-active 类的持续时间为 300ms。而对退出渲染的 DOM 元素，CSSTransition 组件将应用 fade-exit、fade-exit-active 与 fade-exit-done 的 CSS 类，其中 fade-exit、fade-exit-active 类的持续时间同样为 300ms。因而，可对各个 CSS 类设置动画，如在 fade-enter 时设置透明度为 0，在 fade-enter-active 时再设置透明度为 1，并应用 CSS 的 transition 动画过渡，使得每个路由页面都有一个透明度的渐变效果：

```
.fade-enter {
  opacity: 0;
  z-index: 1;
}
.fade-enter.fade-enter-active {
```

```
    opacity: 1;
    transition: opacity 250ms ease-in;
}
```

CSSTransition 中的 timeout 为 300ms，也为 TransitionGroup 缓存子组件（即将被销毁的 CSSTransition 子组件）的时间，被缓存的页面在停留 300ms 后将被 TransitionGroup 销毁。

注意，CSSTransition 的 key 为 location.key，且 Switch 传入了 location，这是因为 TransitionGroup 需要通过不同的 key 来记录即将被销毁的组件元素与即将挂载的组件元素。例如<CSSTransition key="1"/>与<CSSTransition key="2"/>，父组件 TransitionGroup 在路由动画过程中同时将两者保存下来以便实施动画。为 Switch 传入 location 的目的主要是缓存路由切换前的 location，使得离开页面能得到一定程度的保留。在上例中，初始 location 的 pathname 为/a，在路由动画过程中，由于 TransitionGroup 使用 React.cloneElement 进行了元素拷贝，Switch 存在两份，分别为<Switch location={{pathname:'/a'}}/>与<Switch location={{pathname:'/b'}}/>，这样使得路径为/a 的 Route 与路径为/b 的 Route 在动画过程中都得以存在。

如果不使用 CSSTransition，则传入不同的 key，TransitionGroup 将无法区分新旧 CSSTransition 的销毁与挂载。同样，若 Switch 不传入 location，则 location 将从 Context 中获取。虽然 React.cloneElement 拷贝了两份 Switch，但是在渲染时若 location 都从 Context 中获得，由于 Context 中的 location 是同一个值，原 location 没有缓存，则两份 Switch 将渲染同一个新跳转页面，这时若有退出动画，则退出动画将不生效。

对于使用 react-transition-group 库进行路由动画，也可以单独使用 CSSTransition 进行动画的人工触发，无须使用 TransitionGroup 进行管理，如下所示：

```
<Route
path={path}
children={({match}) => {
    return (
        <CSSTransition
            // 人工触发动画
            in={match !== null}
            timeout={300}
            // CSS 动画
            classNames="page"
            // 当 in 为 false 时，在动画退出后，销毁组件
            unmountOnExit>
            <Component />
```

```
            </CSSTransition>
        );
    }}
/>;
```

结合 Route 的 children 属性，可在导航中得知 Route 是否匹配，再通过匹配结果控制 CSSTransition 组件的 in 属性。当 in 为 true 时，触发 enter、entering、entered 的动画生命周期，相应地会为 DOM 元素加上 page-enter、page-enter-active、page-enter-done 的 CSS 类。当 in 为 false 时，触发 exit、exiting、exited 的动画生命周期，同样会为退出的 DOM 元素添加 page-exit、page-exit-active、page-exit-done 的 CSS 类。退出动画的组件与进入动画的组件都将存在 300ms。相应地，匹配命中的组件，即进入动画的组件，从 entering 阶段到 entered 阶段持续 300ms，page-enter、page-enter-active 的 CSS 类会在进入的 DOM 元素中停留 300ms；退出动画的组件从 exiting 阶段到 exited 阶段持续 300ms，page-exit、page-exit-active 的 CSS 类也会持续 300ms。综上所述，可应用 CSS 的进入和退出动画：

```
.page {
    position: absolute;
    left: 15px;
    right: 15px;
}
.page-enter {
    opacity: 0;
    transform: scale(1.1);
}
.page-enter-active {
    opacity: 1;
    transform: scale(1);
    transition: opacity 300ms, transform 300ms;
}
.page-exit {
    opacity: 1;
    transform: scale(1);
}
.page-exit-active {
    opacity: 0;
    transform: scale(0.9);
    transition: opacity 300ms, transform 300ms;
}
```

这样，对于匹配命中的 Route 和没有匹配命中的 Route，都加入了 CSS 动画进行过渡。

在上例 CSSTransition 中，使用了 unmountOnExit 属性，即当 in（in 为触发动画开关）为 false 时，在退出动画执行完成并触发动画结束 exited 生命周期后，销毁 CSSTransition 的子组件。这里对退出动画后的业务组件进行了销毁。

在使用 react-transition-group 库进行路由 CSS 动画过渡时，有一点需要注意：如果 CSS 动画有 transition 的过渡时间，那么一般情况下 CSS 中的动画时间需要小于或等于 CSSTransition 组件的 timeout 属性对应的时间。原因在于 CSSTransition 组件不知晓 CSS 动画的过渡时间。以本节第二个示例进行说明，示例中 timeout 为 300ms，即在路由切换过程中即将被销毁的组件与即将挂载的组件两者同时存在的时间。若从路径/a 跳转到路径/b，则路径为/a 的 Route 与路径为/b 的 Route 同时存在 300ms。当这两个组件同时存在时，路径为/a 的 Route 应用退出的 CSS 动画，路径为/b 的 Route 应用进入的 CSS 动画，但是真正的过渡发生在 CSS 动画上，时间由 CSS 的 transition 控制。300ms 之后，路径为/a 的 Route 页面将被 CSSTransition 销毁，如果 CSS 动画的过渡时间大于 timeout 属性对应的时间，那么 CSS 动画还没执行完毕，timeout 属性对应的时间一到，CSSTransition 就将销毁即将退出的组件，这时由于 CSS 动画还在进行（CSS 动画的过渡时间大于 timeout 属性对应的时间），动画会突然消失，造成非预期结果。

8.6.2 Prompt 组件

可在某些 Form 表单场景中使用 Prompt，获得跳转确认的能力：

```
function Form() {
  const [blocking, setBlocking] = React.useState(false);
  return (
    <form
      onSubmit={event => {
        event.preventDefault();
        /*
          1.获取表单数据
          2.校验通过
          3.提交
          4.提交成功
          5.清空输入
        */
        setBlocking(false);
```

```
        }}
      >
        <Prompt
          when={blocking}
          message={location =>
            `您有输入项未提交或未提交成功,是否确定前往 ${location.pathname}`
          }
        />
        {/* 各表单 */}
        {/* …… */}
        <input
          onChange={event => {
            setBlocking(event.target.value.length > 0);
          }}
        />
        {/* …… */}
        {/* 各表单 */}
        <button>提交表单</button>
      </form>
    );
}
```

Prompt 常用于与未完成的表单结合,常出现在发生跳转、提示用户进行保存等场景中。当需要在跳转前提示用户进行确认时,设置 when 为 true,此时当 message 为字符串或返回字符串的函数时,将弹出对话框提示用户,而跳转与否取决于用户的选择。

使用一个 blocking 状态指明是否应提示用户。当输入框有值时,设置 blocking 为 true,setBlocking(event.target.value.length > 0),这时 Prompt 收到 when 为 true,会调用 history.block 激活跳转前确认。所有的跳转在跳转前都会调用一次 message 方法,由于该方法返回了字符串,因此进行了默认的 window.confirm 弹窗确认(仅 BrowserRouter 或 HashRouter,MemoryRouter 无默认的 getUserConfirmation 方法),如图 8-1 所示。

图 8-1 跳转前弹出的提示框

在单击"确定"按钮后,会进行导航跳转,而单击"取消"按钮,导航行为将被撤回。

如果提交操作完成之后调用 setBlocking(false),则 when 为 false,这时 Prompt 组件会销毁并注销 history.block 的注册,跳转前的确认弹窗将被取消。

如果希望自行实现弹窗提示,则可在创建 history 时传入 getUserConfirmation 函数:

```
// message 为 Prompt 组件上的 message 或者 message 函数返回的字符串
function getUserConfirmation(message, callback) {
  Modal.confirm({
    title: "跳转确认",
    content: message,
    // 允许用户跳转
    onOk: () => callback(true),
    // 不允许跳转
    onCancel: () => callback(false)
  });
}
```

如果不希望有弹窗提示,强制阻止跳转,则可设置 when 为 true,且 message 为 false 或者返回 false 的函数:

```
<Prompt
 when={true}
 message={nextLocation => false}
/>
```

当渲染了如上例中的 Prompt 后,任何单击导航的操作都将被阻止且无提示。若要解除阻止,则可设置 when 为 false,或者 message 函数返回非 false 且非 string 的值,如 true 和 undefine 等。

事实上,由于 Prompt 组件依赖 history.block 的能力,而 history.block 存在刷新丢失等问题(关于 history.block 的问题,可查看 2.6.1 节),因此在使用 Prompt 组件时需要注意。

8.6.3　404 页面

1. 原路径型 404

原路径型 404 页面意味着当访问某个错误页面路径时,URL 路径地址不会发生变化,但是页面内容为未找到页面内容等提示信息:

```
import {
  BrowserRouter as Router,
  Route,
  Switch,
  Redirect
} from "react-router-dom";
export default function() {
  return (
    <Router>
      <Switch>
        <Route path="/" exact component={Home} />
        <Route path="/baz" exact component={Baz} />
        <Route path="/foo" component={Foo} />
        {/* 当以上路径都未命中时，会命中此无 path 的 Route
            原因在于无 path 的 Route 会使用上下文中的 match，这里的上下文由 Router 提供
            Router 提供的上下文中的 match 不为 null，因而一定命中此路由
          */}
        <Route component={NoMatch} />
      </Switch>
    </Router>
  );
}
```

这里利用了 2 个特性：
- Switch 仅渲染第一个命中的 Route。
- 无 path 的 Route 会使用上下文中的 match 作为自身的 match。

基于以上 2 个特性，虽然<Route component={NoMatch} />继承的 Router 的 match 不为 null，一定是匹配命中的，但是由于 Switch 仅渲染一个 Route，因此只有当/、/baz、/foo 都没有命中时，才会渲染这个无 path 的 Route；若/、/baz、/foo 三者中有一个 Route 命中，则仅渲染这个命中的 Route。

2. 重定向型 404

"重定向型 404"意味着当访问没有定义的页面路径时，会重定向跳转到如/404 路径这样的页面，并在此页面展示相关内容。

```
export default function() {
  return (
```

```jsx
<BrowserRouter>
  <Switch>
    <Route path="/" exact component={Home} />
    <Route path="/baz" exact component={Baz} />
    <Route path="/foo" component={Foo} />
    {/*
      定义一个/404 路由页面
    */}
    <Route path="/404" component={NoMatch} />
    {/*
      当/、/baz、/foo、/404 都未命中时会重定向到/404 页面
    */}
    <Redirect to="/404" />
  </Switch>
</BrowserRouter>
);
}
```

相对原路径型 404，重定向型 404 仅需多声明一个/404 的路径，并且同样保证 Redirect 位于 Switch 最后组件的位置。要注意，当命中 Redirect，并跳转到/404 页面后，Redirect 会被销毁，这是由 Switch 仅渲染一个子组件的特性决定的。此时若再访问某个未声明的路径，则 Redirect 组件又会重新实例化。

3. 嵌套型 404

若一级路径有意义，在未匹配命中时，则希望保留一级路径的内容。例如当前有路径/foo/baz，其中/foo 为父路由，/baz 为子路由。若未声明/baz 这个子路由，又希望保留父路由/foo 的内容，则可在子路由中声明：

```jsx
export default function SubComponent(props) {
  return (
    <Switch>
      <Route path=`${props.match.path}/list` component={List} />
      <Route component={SubNoMatch} />
    </Switch>
  );
}
```

同样利用了继承父级 Route 的 match 的特性与 Switch 单一渲染特性，这在一些布局场景中，如页面顶部和左侧都有导航菜单时较为常用。

8.6.4 不销毁未命中路径组件的扩展 Switch

在 6.7.5 节中，曾介绍过缓存 Route。缓存 Route 的一大功能是将 DOM 缓存，区别于原生 Route，缓存 Route 在路由未命中时，不销毁 DOM。但若将缓存 Route 与 Switch 结合，则会造成无法缓存的问题，原因在于原生 Switch 组件仅渲染一个命中子组件，其余未命中子组件都将被销毁，这样应该缓存的 Route 在 Switch 的控制下无法得到有效缓存。基于此种原因，本节为缓存 Route 提供扩展 Switch 组件。扩展 Switch 组件在渲染一个子组件的同时，也能缓存其余曾经渲染过的组件。通过以上分析，设计的 Switch 需要满足如下目标：

- 与原生 Switch 一致，在视图上，Switch 仅选择展示第一个命中的子组件。
- 如果 Switch 的子组件 CacheRoute 曾经命中渲染过，则 CacheRoute 所渲染的组件应该缓存。
- Switch 仅缓存曾经渲染成功的 CacheRoute 组件，从未渲染成功的 CacheRoute 组件不会被缓存。

设计 CacheSwitch，也应与原生 Switch 对应，在使用上与 Switch 一致：

```
function App() {
  return (
    <CacheSwitch>
      <CacheRoute path="/a" component={A} />
      <CacheRoute path="/b" component={B} />
      <CacheRoute path="/c" component={C} />
      <CacheRoute path="/d" component={D} />
    </CacheSwitch>
  );
}
```

参考 8.1.2 节 Switch 源码的实现，从 react-router 包中引入 matchPath、__RouterContext 等，重新实现能缓存 Route 的 Switch：

```
import * as React from "react";
import { __RouterContext, matchPath, SwitchProps } from "react-router";
import invariant from "tiny-invariant";
const RouterContext = __RouterContext;
interface SwitchPropsExt extends SwitchProps {
  // 不缓存模式
  noCache?: boolean;
}
```

```typescript
function CacheSwitch(props: SwitchPropsExt) {
  // 需要一个数组记录哪些子组件曾经渲染过
  const renderedComponentsRef = React.useRef<string[]>([]);
  return (
    <RouterContext.Consumer>
      {context => {
        const location = props.location || context.location;
        // 用于保存渲染组件的数组，由于 CacheSwitch 可能渲染多个子组件，
        // 不像 Switch 仅渲染一个子组件，因此 CacheSwitch 需要缓存子组件
        // 通过数组保存所有需要渲染的子组件
        const components = [];
        let isMatched = false;
        React.Children.forEach(props.children, child => {
          // 还原 Switch 的行为，在命中后不再进行后续 child 的操作
          if (props.noCache && isMatched) {
            return;
          }
          if (React.isValidElement(child)) {
            const element = child;
            // 获取 Route 的 path
            const path = child.props.path || child.props.from;
            // 与 Switch 的 match 一致
            const match = path
              ? matchPath(location.pathname, { ...child.props, path })
              : context.match;
            const compnentIdentity = element.key || path;
            // 渲染组件方法，通过判断命中数组确保组件命中过
            const renderComponent = forceHide => {
              invariant(compnentIdentity, `请确认组件${element.type}的 key`);
              // 曾经渲染过的组件应该继续渲染，隐藏与否根据 forceHide 参数决定
              renderedComponentsRef.current.includes(compnentIdentity) &&
                components.push(
                  // 使用 cloneElement 保留原 element 的 props
                  React.cloneElement(element, {
                    // 与 Switch 一致，将 Switch 的 location 传入 Route
                    location,
                    // 为 Route 传入 computedMatch，Route 便无须再计算一次命中情况
                    computedMatch: match,
```

```js
          // 在传入 noCache 为 true 时，Route 行为与原生一致
          // 使用 6.7.5 节中的 CacheRoute 需要传入对应字段
          shouldDestroyDomWhenNotMatch: props.noCache,
          // CacheRoute 组件的属性，渲染时 CSS 强制控制
          forceHide: forceHide,
          // 使用 key 值是必要的，因为 Switch 下的 children 是组件数组
          // 优先使用 element 的 key 值，如果没有，则使用 path 值作为 key 值
          key: compnentIdentity
        })
      );
    };
    if (match) {
      if (!isMatched) {
        // 此组件已经满足渲染要求，更新标识符
        !renderedComponentsRef.current.includes(compnentIdentity) &&
          renderedComponentsRef.current.push(compnentIdentity);
        // 第一次匹配成功
        isMatched = true;
        // 第一次匹配成功，不强制隐藏
        renderComponent(false);
      } else {
        // 非第一次匹配成功，强制隐藏
        renderComponent(true);
      }
    } else {
      // 未匹配成功，强制隐藏
      renderComponent(true);
    }
  }
});
// 更新一次标识组件渲染的数组，以满足在 key 变化时，旧的 key 能得到清理
renderedComponentsRef.current = components.map(element => element.key);
return components;
    }}
  </RouterContext.Consumer>
  );
}
export default CacheSwitch;
```

缓存 Switch 的运行原理是渲染多个子组件，但仅选择第一个命中路径的组件进行展示。这样由于多个子组件得到了渲染，各个子组件的内容不会被销毁，只要子组件使用了 CacheRoute，便能实现缓存。

在实现缓存 Switch 时，首先需要声明一个标识数组，用于保存曾经渲染成功的组件。这样当 Switch 再次渲染时，才可再次渲染这些缓存组件。数组中的内容为组件标识符 compnentIdentity，为组件的 key 值或者在没有 key 值时获取组件对应的 path 路径。由于 Switch 可使 Redirect 组件成为条件跳转组件，因此 path 还可以为 Redirect 的 from 属性值。模仿 8.1.2 节的 Switch 源码，在 Switch 每次渲染时，用 isMatched 标识第一次命中匹配。由于 Switch 的特性是仅展示第一个匹配的子组件，因此这时可以调用 renderComponent(false)渲染此组件，false 参数表示不强制隐藏，且由于第一次命中的子组件将得到展示，这里将更新 renderedComponentsRef 对应的数组，入栈该组件的标识符，表示该组件是真正渲染过的组件。如果真正渲染过的组件标识符（组件的 key 值或者 path）在渲染过程中发生了变化，则可以在搜集完渲染组件后，更新一次 renderedComponentsRef 对应的数组，以满足在 key 变化时，旧的 key 能得到清理。如果在组件第一次匹配渲染后，后续也有组件匹配，且曾经渲染过（renderedComponentsRef 包含组件标识符），则调用 renderComponent(true)强制 CSS 隐藏，缓存组件。

在操作子组件时，在使用 cloneElement 保留原 element 子组件的 props 的同时，也如同原生 Switch 组件一样，将 location、computedMatch 的值传入子组件中，使得子组件不必重新计算 match。除此之外，在调用 React.cloneElement 方法时为子组件增加 forceHide、shouldDestroyDomWhenNotMatch 等 CacheRoute 的属性（见 6.7.5 节）。同时，由于渲染多个子组件，各子组件存在于一个数组中，因此需要为各子组件设置不同的 key 值。key 值在选取上优先使用原 element 的 key 值，若没有声明 key，则可以取 path 值作为 key 值，并通过 invariant 强制保证 key 值，以保证 React 组件数组 key 的完整性。

同时，组件还支持不缓存属性 noCache。通过此属性，Switch 仅保留第一个命中路径的子组件，并且在渲染该命中子组件时，也将 props.noCache 赋值到 shouldDestroyDomWhenNotMatch，促使 CacheRoute 在相关场景中销毁组件。

8.7 小结

本章主要介绍了 React Router 中的 Switch、Redirect、Prompt、matchPath 等组件或方法。Switch

作为路由选择器，在同级多 Route 匹配场景下使用可保证单个 Route 的渲染。Redirect 作为重定向组件，提供了便捷的重定向导航。使用 Prompt 可对用户导航行为进行控制，如弹窗询问等。matchPath 方法使得开发者可便捷、高效地人工判断路径匹配情况。通过本章中源码及相关示例的介绍，既能加深对 React Router 的理解，也能深入学习及掌握各组件的设计思路及实际用法，更能提升开发效率和代码质量，为能更便捷、高效地开发 React 应用打好基础。

参考文献

[1] https://github.com/ReactTraining/react-router.

[2] https://reacttraining.com/react-router/web/example/animated-transitions.

[3] http://reactcommunity.org/react-transition-group/with-react-router.

第 9 章 进阶实战案例

9.1 路由组件的滚动恢复

在实际开发中，往往会碰到一些场景，如当鼠标滚轮滚动到一定位置的列表时，单击一个具体的列表项，就跳到了这个列表项的详情页；当导航返回时，为了保持良好的用户体验，希望在回到列表时，还能回到之前鼠标滚轮滚动到的位置。

9.1.1 scrollRestoration

在 Chrome 46 之后，history 引入了 scrollRestoration 属性。该属性提供两个值。第一个是 auto，为默认值，基于元素进行位置记录，浏览器会原生地记录下 window 中某个元素的滚动位置。不管是浏览器强制刷新切换页面，还是使用 pushState、replaceState 方法改变页面状态，都可能由于一些操作使滚动条位置发生变化，而这个属性能始终让元素恢复到之前的屏幕范围内。但是要注意，其只能记录下在 window 中滚动的元素，如果是某个容器中的局部滚动，则浏览器是无法识别出的。事实上，元素在某个容器或容器的容器中时，浏览器并不知道开发者想要保存哪个 DOM 节点的滚动位置，所以在这种情况下会失效。IE 与 Safari 目前也不支持这个属性。对于另一个值 manual，其等于把属性设置为手动进行，这将丢失上述原生的恢复能力，在浏览器强制刷新切换页面，或者使用 pushState、replaceState 方法改变页面状态时，滚动条都会回到顶部。

9.1.2 容器元素滚动恢复

实现容器元素滚动恢复，具体的思路是，在路由切换、元素即将消失于屏幕前，记录下元素的滚动位置，当元素重新渲染或出现于屏幕中时，再恢复这个元素的滚动位置。得益于 React Router 的设计思路，类似 Router 组件负责搜集 location 变化，并把状态向下传递，设计滚动管理组件 ScrollManager，用于管理整个应用的滚动状态。同理，类似于 React Router 中的 Route，作为具体执行者进行路由匹配，设计对应的滚动恢复执行者 ScrollElement，用以执行具体的恢复逻辑。本节示例使用 React v16.8。

9.1.3 滚动管理者 ScrollManager

滚动管理者作为整个应用的管理员，应该具有一个管理者对象，用来设置原始滚动位置、恢复和保存原始节点等，通过 React 的 Context，将该对象分发给具体的滚动恢复执行者。

```
export interface IManager {
  registerOrUpdateNode: (key: string, node: HTMLElement) => void;
  setLocation: (key: string, node: HTMLElement | null) => void;
  setMatch: (key: string, matched: boolean) => void;
  restoreLocation: (key: string) => void;
  unRegisterNode: (key: string) => void;
}
```

上述 manager 对象拥有注册 HTMLElement 元素 registerOrUpdateNode，设置 HTMLElement 元素位置 setLocation、restoreLocation 恢复位置等方法，但是缺少了缓存对象。对于缓存可以使用 React.useRef，这个接口类似于类的属性：

```
/*
        注册缓存内存，类似于 this.cache
*/
const locationCache = React.useRef<{
  [key: string]: { x: number; y: number };
}>({});
const nodeCache = React.useRef<{ [key: string]: HTMLElement | null }>({});
const matchCache = React.useRef<{ [key: string]: boolean }>({});
const cancelRestoreFnCache = React.useRef<{ [key: string]: () => void }>({});
```

通过 React.useRef 设置了各类缓存。接下来，来实现 manager 对象。

实现 manager 对象要用到上述的缓存对象，并使用 key 作为缓存的索引。关于 key，会在 scrollElement 中进行说明。

```
const manager = {
  registerOrUpdateNode: (key: string, node: HTMLElement) => {
    nodeCache.current[key] = node;
  },

  setMatch: (key: string, matched: boolean) => {
    matchCache.current[key] = matched;
  },

  unRegisterNode: (key: string) => {
    nodeCache.current[key] = null;
  },

  setLocation: (key: string, node: HTMLElement | null) => {
    if (!node) {
      return;
    }
    locationCache.current[key] = { x: node.scrollLeft, y: node.scrollTop };
  },

  restoreLocation: (key: string) => {
    if (!locationCache.current[key]) {
      return;
    }
    const { x, y } = locationCache.current[key];
    nodeCache.current[key].scrollLeft = x;
    nodeCache.current[key].scrollTop = y;
  }
};
```

其中，registerOrUpdateNode 用于保存当前的真实 DOM 节点，unRegisterNode 用于清空节点缓存，setLocation 用于保存页面切换前的滚动位置，restoreLocation 用于恢复。在简单实现了 manager 对象之后，便可以通过 Context 将 manager 对象向下传递：

```
<ScrollManagerContext.Provider value={manager}>
  {shouldChild && props.children}
</ScrollManagerContext.Provider>
```

这样，一个基本的 ScrollManager 组件雏形就完成了。但 manager 还需要一个重要的能力：获知元素导航切换前的位置。只有实现了这个能力，manager 才能保存页面切换前的滚动位置。

接下来，需要实现 ScrollManager 组件以获知元素导航切换前的位置。

在 React Router 中使用了 props.history.listen，一切路由状态的切换都从 props.history.listen 中发起。由于 listen 可以监听多个函数，因此便可利用 props.history.listen。在 React Router 路由状态切换前，插入一段监听函数，获得相关的节点信息。在获得变化前的节点信息之后，再执行 React Router 的路由切换。路径为：

```
loactionChange---->getDomLocation----->路由 update
```

示例中使用了一个状态 shouldChild，来确保监听函数在触发顺序上一定是先于 React Router 中 Router 的监听函数的，也可以选择在使用 useState 时，传入的第一个参数为函数，并在其中执行构造函数的逻辑。在实现上，使用 useEffect 模拟了 componentDidMount 和 componentWillUnmount。在 history.listen 回调函数中，会对 nodeCache 中的每个变量判断 matchCache。若 matchCache 为 true，则表明从当前 match（路由渲染的页面）离开。所以离开之前，保存 scroll 位置：

```
useEffect(() => {
    const unlisten = props.history.listen((_location, _action) => {
      // 每次 location 变化时，保存节点信息
      // 这个回调要在 history 的所有回调中第一个执行，原因是这时还没进行
  // setState,并且即将要进行 setState,在这个回调中得到的状态或 dom 属
  //性是进行状态更新前的
      const cacheNodes = Object.entries(nodeCache.current);
      cacheNodes.forEach(entry => {
        const [key, node] = entry;
        // matchCache 为 true，表明从当前 match(路由渲染的页面)离开
        if (matchCache.current[key]) {
          manager.setLocation(key, node);
        }
      });
    });
    // 保证先监听完上面的回调函数后，才实例化 Router
  // 保证了上面的回调函数最先入栈
    setShouldChild(true);
    return () => {
      // 销毁时清空所有内存信息
```

```
        locationCache.current = {};
        matchCache.current = {};
        nodeCache.current = {};
        cancelRestoreFnCache.current = {};
        Object.values(cancelRestoreFnCache.current).forEach(
          cancel => cancel && cancel()
        );
        unlisten();
      };
      // 依赖为空，仅在挂载和销毁时调用
    }, []);
```

在销毁组件时，要清空所有的缓存，防止内存泄露。ScrollManager 在使用时放在 Router 的外侧，这样可以控制 Router 的实例化：

```
<ScrollManager history={history}>
  <Router history={history}>.......... ..........</Router>
</ScrollManager>;
```

9.1.4 滚动恢复执行者 ScrollElement

ScrollElement 的主要职责是控制真实的 HTMLElement 元素，决定缓存的 key，包括决定何时触发恢复、何时保存原始 HTMLElement 的引用、设置是否需要保存位置的标志等。ScrollElement 接收如下 props：

```
interface IProps {
  // 必须缓存的 key
  scrollKey: string;
  children?: React.ReactElement;
  // 为 true 时触发滚动恢复
  when?: boolean;
  // 外部传入 Ref
  getRef?: () => HTMLElement;
}
```

其中 scrollKey 为必须传入的字段，用来标志缓存的具体元素、缓存的位置信息、缓存的状态等，需要全局唯一。使用 when 字段可控制是否需要进行滚动恢复。ScrollElement 本质上是一个代理，会拿到子元素的 Ref，接管其控制权。也可以自行实现 getRef 传入组件中，组件会对传入的 Ref 进行操作。

```ts
import * as React from "react";
import { useContext, useLayoutEffect, useEffect } from "react";
import { __RouterContext as RouterContext, match } from "react-router";
import { IManager, ScrollManagerContext } from "./scroll-manager";
// ScrollElement
export default function(props: IProps) {
  const nodeRef = React.useRef<HTMLElement>();
  const manager: IManager = useContext<IManager>(ScrollManagerContext);
  const currentMatch = useContext(RouterContext).match;
  useEffect(() => {
    const handler = function(event: Event) {
      if (nodeRef.current === event.target) {
        manager.setLocation(props.scrollKey, nodeRef.current);
      }
    };
    // 在 window 上监听 scroll 事件, 获取 scroll 事件触发 target, 并更新位置
    window.addEventListener("scroll", handler, true);
    // 移除事件
    return () => window.removeEventListener("scroll", handler, true);
  }, [props.scrollKey]);
  useLayoutEffect(() => {
    if (props.getRef) {
        nodeRef.current = props.getRef();
    }
    if (currentMatch) {
      // 设置标志, 表明在 location 改变时, 可以保存路径
      manager.setMatch(props.scrollKey, true);
      // 更新 Ref, 代理的 DOM 可能会重新挂载, 所以要每次更新
      nodeRef.current &&
        manager.registerOrUpdateNode(props.scrollKey, nodeRef.current);
      // 恢复原先滑动过的位置, 可通过外部 props 通知是否需要进行恢复,
      // 一般为:when={xxx.length>0}
      (props.when === undefined || props.when) &&
        manager.restoreLocation(props.scrollKey);
    } else {
      // 没命中设置标志, 不保存路径
      manager.setMatch(props.scrollKey, false);
    }
```

```
    // 销毁时注销这个 node
    return () => manager.unRegisterNode(props.scrollKey);
  });
  if (props.getRef) {
    return props.children;
  }
  const onlyOneChild = React.Children.only(props.children);
  // 代理第一个 child，需要是真实的 dom、div、h1、h2……不能是组件
  if (typeof onlyOneChild.type === "string") {
    // 必须是原生 tag 在合格的子元素上，加上新的 Ref
    // 以便接管控制权
    return React.cloneElement(onlyOneChild, { ref: nodeRef });
  } else {
    console.warn(
      "-------------滚动恢复将失效，ScrollElement 的 children 必须为原生的单个 HTML 标签-------------"
    );
    return props.children;
  }
}
```

在上例中，之所以使用 useLayoutEffect，是为了在浏览器绘制前执行相关的 DOM 变更方法。浏览器可在绘制阶段统一处理 DOM 变更，从而避免导致页面的闪动。上例使用 useLayoutEffect，会执行 componentDidMount、componentDidUpdate、componentWillUnmount 生命周期。在初次加载或每次更新时，会根据当前的 Route 匹配与否进行对应的处理。如果 Route 匹配成功，则表明当前的 ScrollElement 组件应是渲染的，这时可以在 effect 中执行更新 Ref 的操作。之所以在 effect 中执行更新，是因为代理的 DOM 可能会变化，所以要每次更新。同时还需要设置标志，表明在 location 改变时，是可以保存滚动位置的，相当于告诉 manager，node 此刻渲染成功了，可以在离开页面时把现在的位置保留下来。如 match 为 false，则表明此刻组件并没有跟路由匹配上，不应渲染，所以 manager 此刻也不应保存这个元素的位置信息。在元素匹配成功，并且更新了 DOM 后，便可在 effect 中恢复元素到原来的位置：

```
(props.when === undefined || props.when) &&
    manager.restoreLocation(props.scrollKey);
```

在之前的 manager 部分有过介绍，这时候会根据 key 获得缓存的位置信息，并设置 DOM 属性，以恢复元素的位置：

```
restoreLocation: (key: string) => {
  if (!locationCache.current[key]) {
    return;
  }
  const { x, y } = locationCache.current[key];
  nodeCache.current[key].scrollLeft = x;
  nodeCache.current[key].scrollTop = y;
}
```

元素的恢复还可以通过 when 属性来控制，当 when 为 true 时执行恢复，如下所示：

```
<ScrollElement
  when={bigArray.length > 0}
  scrollKey="xxxxx(全局唯一)"
>
  <ul>.......... ..........</ul>
</ScrollElement>
```

注意，如果 ScrollElement 是第一次渲染，由于没有保存过滚动位置，因此在 effect 中执行 restoreLocation 不会触发任何行为。

9.1.5 多次尝试机制

在上面的恢复过程中，只执行了一次恢复行为：

```
nodeCache.current[key].scrollLeft = x;
nodeCache.current[key].scrollTop = y;
```

对于一些浏览器，有可能执行一次位置赋值后浏览器得到的结果并不如预期，可能会有偏差。这时可引入一个工具函数，使得可以多次执行某方法：

```
// 可取消的重复尝试函数
const tryMutilTimes = (
  callback: (...args: any[]) => void,
  tickInterval: number,
  timeout: number
) => {
  const timeId = setInterval(callback, tickInterval);
  setTimeout(() => {
    clearTimeout(timeId);
  }, timeout);
```

```
    return () => clearTimeout(timeId);
};
```

使用一个定时器多次执行 callback，同时设置一个执行时间上限，并返回一个取消函数给外部。tryMutilTimes 为可取消的，这给 restoreLocation 提供了很好的控制能力，更改后的 restoreLocation 如下：

```
restoreLocation: (key: string) => {
    if (!locationCache.current[key]) {
      return;
    }
    const { x, y } = locationCache.current[key];
    let shoudNextTick = true;
    cancelRestoreFnCache.current[key] = tryMutilTimes(
      () => {
        if (shoudNextTick && nodeCache.current[key]) {
          nodeCache.current[key]!.scrollLeft = x;
          nodeCache.current[key]!.scrollTop = y;
          // 如果恢复成功，就取消，不用再恢复了
          if (
            nodeCache.current[key]!.scrollTop === y &&
            nodeCache.current[key]!.scrollLeft === x
          ) {
            shoudNextTick = false;
            cancelRestoreFnCache.current[key]();
          }
        }
        // 每隔 50ms 尝试一次恢复，尝试到 500ms 结束，时间可配置
      },
      props.restoreInterval || 50,
      props.tryRestoreTimeout || 500
    );
}
```

设置一个时间间隔，多次尝试滚动恢复的操作，如果最终恢复的位置与预期一致，则可取消 tryMutilTimes 多次尝试，滚动恢复结束。如果与预期不符，则再次尝试，直到在 timeout 时间内与预期的滚动位置一致。

9.2 异步 history 方法

在以往使用 history.push 或 history.replace 方法时，有时发起跳转的页面在跳转后未被销毁（如 6.7.5 节的缓存 Route）；同时，需要在跳转成功后清理原页面的状态，如原页面的一个弹窗状态、复选框的勾选状态，以往的清理办法多为在 componentWillUnMount 中，或者监听到页面离开后在 onLeave 中（将在 9.4 节介绍）进行状态的清除。下面介绍一种 history 的实现方法。

在实现上，提升 history 的能力，使得 history.push 或 history.replace 等方法在调用后返回一个 promise。当 promise 解决后，则可认为页面已成功完成了跳转。使用方式如下：

```
<>
  <Modal show={this.state.show} />
  {/* ………… */}
  <a
    onClick={async () => {
      this.setState({ show: true });
      // 返回 promise
      await this.props.history.push(
        `/user/${id}`
      );
      this.setState({ show: false });
    }}
  >
    查看
  </a>
</>
```

this.props.history.push 返回了 promise，当页面成功跳转，完成下一个页面的渲染后，选择将弹窗状态设置为 false。

要实现以上功能，需要引入：
- 能返回 promise 的 history 的各种跳转方法。
- 能获知导航过程中下一个页面是否成功完成跳转的顶层组件。

9.2.1 提升 history 方法

原始的 history 可以在各级组件中通过上下文传递，因而只用提升原始 history 的各方法，在保持原有 history 方法行为不变的基础之上，引入新的能力：

```
export function enhanceHistoryNavigation(
  history: IHistory,
  methods = ["go", "goBack", "goForward", "push", "replace"]
) {
  const originHistoryMethods = {};
  methods.forEach(method => {
    originHistoryMethods[method] = history[method];
  });
  let transitioning = false;
  let transitionEndResolver = () => {};
  methods.forEach(method => {
    history[method] = (...args: any[]) => {
      if (transitioning) {
      // 如果页面跳转还未完成，则所有跳转仅进行 replace 操作
        originHistoryMethods["replace"](...args);
      } else {
        originHistoryMethods[method](...args);
      }
      transitioning = true;
      transitionEndResolver = null;
    // 返回待解决的 promise，保存 resolve 函数
      return new Promise(resolve => {
        transitionEndResolver = resolve;
      });
    };
  });
  history._onEnd = () => {
    transitioning = false;
  // 提供私有解决方法
    transitionEndResolver && transitionEndResolver();
  };
  return history;
}
```

引入 enhanceHistoryNavigation 方法，将 history 作为参数传入。在初始时，将原始 history 的各方法保留到 originHistoryMethods 中，再对各 history 方法进行扩写。在扩写 history 方法时，在

调用完原始方法后，返回一个未解决的 promise，并将其 resolve 函数保存到 transitionEndResolver 变量中。同时，history 提供一个内部方法_onEnd，在_onEnd 内部会把导航过程状态 transitioning 设置为 false，并且调用 transitionEndResolver 更新 promise 的状态。_onEnd 的作用是提供给能感知页面变化的顶层组件使用，由顶层组件决定具体的调用时机。

同时引入 transitioning 变量，用于表明导航过程是否处于进行中。由于是异步的导航切换方法，因此可用 transitioning 标识导航是否结束。如果 transitioning 为 true，导航还未结束，则再次调用 history.push 方法，此种情况会使用 history.replace 方法进行替代，保证栈内的浏览记录数量。在每次跳转返回 promise 前，都会设置 transitioning 为 true，且在_onEnd 内设置为 false。

9.2.2 导航感知

在设计顶层组件时，需要获知导航过程中下一个页面是否成功跳转。可选用 effect 方法，在 effect 中进行判断。设计顶层监听组件 NavigationListener，其在 Router 组件之下引入：

```
const history = createBrowserHistory();
const enhancedHistory = enhanceHistoryNavigation(
  history
);
function App() {
  return (
    <div className="App">
      <Router history={enhancedHistory}>
        {/* 导航监听 */}
        <NavigationListener>
          <Route path="/a" />
          ……
          <Route path="/b" />
        </NavigationListener>
      </Router>
    </div>
  );
}
```

实现 NavigationListener 组件，利用 effect 监听每次 location 变化后 DOM 的更新，同时使用 previousLocationRef 记录上一次的 location 信息。之所以选择在 effect 中进行处理，是因为 effect 触发时表明页面已经跳转成功，下一个页面的初始 DOM 已经准备就绪。由于组件跳转都在

NavigationListener 的 update 阶段，因此可以使用 useRef 产生的 mounted 变量忽略 NavigationListener 组件 componentDidMount 的情况，在 effect 中仅处理组件的 update。

为了获取当前地址，使用 useContext 获得当前 Router 上下文中的 history 与 location 信息。history 信息既可以选择从上下文中获得，也可以选择从外部 props 传入，两者效果一致。在 effect 中，选择 history 库的 locationsAreEqual 方法，用以判断当前 location 与上一次 location 是否一致。若不一致，则表明新页面已经跳转成功，且新页面初始 DOM 已经准备就绪。这时可以触发第一步中挂载到 history 上的 history._onEnd 方法，将导航设置为结束状态，_onEnd 将调用 history.push 等方法所返回的 promise 内的 resolve 方法（通过变量 transitionEndResolver 保存），使得 history.push 或 history.replace 等方法返回的 promise 得到解决。NavigationListener 组件如下：

```
import { __RouterContext as RouterContext } from "react-router";
import * as React from "react";
import { useContext, useEffect } from "react";
import { History, locationsAreEqual } from "history";
interface IHistory extends History {
  _onEnd?: () => void;
}
export default function(props) {
  const history: IHistory = useContext(RouterContext).history;
  const location = useContext(RouterContext).location;
  const previousLocationRef = React.useRef({});
  const previousLocation = previousLocationRef.current;
  const mounted = React.useRef(false);
  // useEffect，说明新跳转页面已经挂载
  useEffect(() => {
    if (!mounted.current) {
    // 页面跳转过程都在 didUpdate 节点中进行，通过 mounted 忽略组件的初始挂载
      mounted.current = true;
    } else if (!locationsAreEqual(location, previousLocation)) {
      // 若在 effect 中不相等，则调用_onEnd 方法
      history._onEnd && history._onEnd();
    }
    previousLocationRef.current = location;
  });
  return props.children;
}
```

9.3 为路由引入 hash 定位

通常在一个页面中使用 hash 来定位一个具体的元素，从而实现锚点的功能。常用的办法是设置一个 a 标签，并指定 href 的形式实现锚点。但是在使用 Link 进行带锚点的跳转时，会遇到两个问题：

- 使用 Link 组件或 NavLink 组件，如<NavLink to="/a#name"/>定位 name 元素时，定位功能会失效，NavLink 渲染的 a 元素并不是一个纯锚点定位元素。
- 页面刷新，也将无法定位到锚点元素。

若仅使用带锚点信息的 a 标签定位到锚点元素，则在实现手段上过于单一。因此，本节将分两部分支持与路由相关的 hash 定位：

- 页面加载，导航切换过程的锚点支持。
- 异步数据加载，组件渲染后的锚点支持。

9.3.1 页面加载

对于第一个支持的功能，思路是在页面挂载或更新后，获取 location 中的 hash 部分，并人工进行元素定位，对应的生命周期为 componentDidMount 和 componentDidUpdate。由于本书使用了 Hooks，本节将使用 effect 来处理元素定位。在第 5 章曾介绍过 Router，顶层的 Router 负责路径的分发，会有一个 setState 操作，所有的 children 应该重新渲染，在 React 组件树中从上到下调用 render 函数渲染后，再从下到上触发 componentDidUpdate 生命周期。利用这一特性，设计一个 Router 的一级子组件，负责获知整个路由系统的 mount 和 update 情况，便可在其中进行锚点定位的支持：

```
// 注意要放到 Router 的子元素下
export default function(props) {
  // 相当于调用了 Context.Consumer，当 context.Provider 的 value 更新时，使用到
  // Context.Consumer 的组件会重新渲染，这里利用了这一特性
  React.useContext(__RouterContext);
  useEffect(() => {
    const { location, action } = props.history;
    // 根据 hash 定位元素，将 history 中的 location 传入
    locationHash(location, action);
  });
```

```
    return props.children;
}
```

有一点需要注意，这里虽然使用了 React.useContext(__RouterContext)，但是没有利用 Context 的返回值，这相当于使用了组件 Context.Consumer。当 Context.Provider 的 value 更新时，使用到 Context.Consumer 的组件会重新渲染，这里利用了这一特性，从而保证 Router 中的 Provider 的值变化时，组件能重新渲染。

上例中组件使用了 useEffect，监听了初始化或路径变化时组件的 componentDidMount 和 componentDidUpdate 生命周期。使用和 Router 一致的 history，得到包括 hash 值的 location 路径，便可进行锚点定位。这里的定位方式如下：

```
// 在一定时间内多次执行 callback
const tryMutilTimes = (
  // eslint-disable-next-line
  callback: (...args: any[]) => void,
  tickInterval: number,
  timeout: number
) => {
  const timeId = setInterval(callback, tickInterval);
  setTimeout(() => {
    clearTimeout(timeId);
  }, timeout);
  return () => clearTimeout(timeId);
};
function locationHash(location, action) {
  // 这里的 location 不区分 BrowserRouter 或 HashRouter，统一都有 hash 值
    if (typeof location.hash === "string" && location.hash.length > 1) {
      const elementId = location.hash.substring(1);
      // 多次尝试机制
      const cancelTry = tryMutilTimes(
        () => {
          // 获取到原生元素
          const nvElement = document.getElementById(elementId);
          if (nvElement) {
            nvElement.scrollIntoView();
            cancelTry();
          }
        },
```

```
            10,
            100
        );
    }
}
```

上例使用了 scrollIntoView 进行定位，并加入了多次尝试机制。

可以通过 POP、PUSH、REPLACE 区别 history 的跳转行为，对应 history.push 方法的 action 为 PUSH，对应 history.replace 方法的 action 为 REPLACE。如果单击浏览器的"前进"或"后退"按钮，则对应 history 的 action 为 POP；如果希望只有页面内的单击触发元素定位，则可以判断 action 为 PUSH 或 REPLACE。这样单击浏览器的"前进"或"后退"按钮，就会忽略掉元素定位的功能。但要注意，通过 window.location.hash 人工进行定位，因为这类定位没有通过 history.push 或 history.replace 方法，所以这在 history 看来触发的 action 将为 POP，会认为是单击浏览器的"前进"或"后退"按钮。

在定位使用上，可使用 Link 或 NavLink，如当前页面为/a，则单击 to 为/b#name 的 Link 后，除了导航到/b 页面，也会定位到 id 为 name 的 DOM 元素。综上所述，ElementLocated 最终使用方式如下：

```
import ElementLocated from "./elementLocation";
<Router history={history}>
  {/*元素定位*/}
  <ElementLocated history={history}>
    <Route />
    ............
    <Route />
  </ElementLocated>
</Router>
```

由于监听了路由系统的变化，无论刷新页面，还是使用 Link 或 NavLink 进行导航切换页面，都处于路由系统的监控范围内，这样便能支持页面加载的场景。

9.3.2 异步数据加载

上述场景都是在路由系统初始化或更新变化时监听并进行处理的，侵入性较小，业务代码或组件无须加入额外代码。但是在真实场景中，数据大多是异步获取并将异步数据渲染到 UI 的，这类在局部范围内的 UI 变化不在 React Router 的控制之下，因为这类 UI 变动并不是 React Router

所引起的，上述的监控也将失效，这就需要引入其他办法。考虑在需要异步定位 UI 的组件外引入一个父组件，这个最外层的父组件可感知组件的异步数据变化，componentDidMount 生命周期会依次从下到上执行，对应 componentDidUpdate 生命周期也会从下到上执行。选择使用 effect 监听变化，并使用 Router 上下文，获取 React 组件树中最近的 Route 中的 Provider 所提供的值，选择在 match 路由匹配命中时才进行定位，避免未命中时的无效定位。父组件设计如下：

```
// 父组件提供 hash 定位支持
export function ElementLocated(props) {
  const router = React.useContext(__RouterContext);
  const { location, match, history } = router;
  // 父组件在 effect 中处理 DOM 的相关操作
  React.useEffect(() => {
    if (match && locationHash) {
      locationHash(location, history.action);
    }
  });
  return props.children;
}
```

这样在使用时仅需要在最外层套上组件：

```
import * as React from "react";
import { ElementLocated } from './elementLocated';
export default function(props) {
  const array = Array.from({ length: 100 });
  const [shouldChild,setChild] = React.useState(false);
  // 模拟异步
  !shouldChild && setTimeout(()=>{
      setChild(true)
  },1000);
  // 使用 ElementLocated
  return (
    <ElementLocated>
    {!shouldChild && <div>Loading........................</div>}
    <ul>
      {shouldChild && array.map((v, i) => {
        return (
          <li key={i} id={`${i}`}>
            {i}
```

```
          </li>
        );
      })}
    </ul>
  </ElementLocated>
);
}
```

这样在异步获取数据后，整个 UI 组件包括新加入的包裹父组件会重新渲染，会从下到上执行 update，从而可在 effect 中，根据 hash 的值进行元素定位。

9.4 为组件引入路由生命周期

9.4.1 路由生命周期

在 React Router v3.x 中，配置化路径支持定义如 onEnter 的路由回调，如：

```
const routes = {
  path: '/',
  component: App,
  indexRoute: { component: Dashboard },
  childRoutes: [
    {
      path: 'inbox',
      component: Inbox,
      childRoutes: [{
        path: 'messages/:id',
        onEnter: ({ params }, replace) => replace(`/messages/${params.id}`)
      }]
    },
  ]
}
render(<Router routes={routes} />, document.body)
```

在 React Router v4.x 之后，移除了 onEnter、onLeave 的接口，由于 React Router 由静态路由改为了动态路由，路由的生命周期与组件的生命周期一致。动态路由的生命周期在组件中，如 componentDidMount 对应 onEnter、componentWillUnmount 对应 onLeave。React Router 不再维护

onEnter、onLeave 生命周期。但是若遇到 6.7.5 节中的 CacheRoute 或单纯保存 DOM 的场景，如：

```
<Route
  path="/list-C/:id"
  children={props => {
    return (
      <div
        style={{
          display: props.match
            ? "block"
            : "none"
        }}
      >
        <Comp {...props} />
      </div>
    );
  }}
/>
```

由于 children 属性无论路由命中与否始终会渲染组件，所以在离开页面时，组件 Comp 并不会触发 componentWillUnmount 生命周期，仅会触发更新的生命周期函数，且接收到的最新的 props.match 为 null。即便当前路由没有匹配命中，组件 Comp 的 render 函数依旧会执行，这与生命周期的对应关系有一定冲突。因此，下面设计一种机制来兼容此种场景。目标是提供统一的生命周期接口兼容 CacheRoute 及 Route 的 render 属性与 children 渲染场景，业务组件可不需要关心路由匹配命中或者未匹配命中时所触发的生命周期，不再关心如 componentDidMount、componentWillUnmount 等。

设计 didEnter 和 onLeave 函数，其中 didEnter 用于注册路由进入回调，回调函数将在已进入相应路由后触发；onLeave 用于注册路由退出回调，回调函数将在离开当前路由时触发。它们的签名为：

```
// 可通过 props.onLeave 注册路由退出回调函数
interface onLeave {
  (cb: onLeaveFuncType): () => void;
}
// 路由退出回调函数的签名
interface onLeaveFuncType {
  /*
    第一个参数标记页面匹配的路由信息，与 6.3 节所述一致
```

```
  第二个参数标记组件离开时 DOM 是否销毁
*/
  (props?: RouteComponentProps, isUnmount?: boolean): void;
}
// 可通过 props.didEnter 注册路由进入回调函数
interface didEnter {
  didEnter(cb: didEnterFuncType): () => void;
}
// 路由进入回调函数的签名
interface didEnterFuncType {
  /*
    第一个参数为进入页面时的路由信息, 与 6.3 节所述一致, 可获取到 match 等值
    第二个参数表示页面切换是否是更新切换, 如一个组件在多个路径都能命中, 一般是
    Route 路径为/user/:id 的场景, 组件在/user/1、/user/2 等路径都能渲染, 并从
    /user/1 切换到 /user/2
  */
  (props?: RouteComponentProps, isUpdate?: boolean): void;
}
```

通过 props 将 didEnter 和 onLeave 注入路由组件中,组件可调用 props.didEnter、props.onLeave 注册路由生命周期回调函数。didEnter 的回调函数接收两个参数, 其签名为 didEnterFuncType。第一个参数为进入页面时的路由信息, 与 6.3 节介绍的 props 一致, 主要包含 match 对象、location 地址对象和 history 对象。第二个参数 isUpdate 标识进入路由的模式, 一般在同一个组件命中多个路由路径时,该参数为 true。例如 Route 路径为/user/:id,从/user/1 进入/user/2 时。同样,onLeave 的回调函数也接收 2 个参数, 其签名为 onLeaveFuncType。第一个参数为退出页面时的路由信息, 与 RouteComponentProps 一致。第二个参数 isUnmount 标记页面离开时的模式, 如果为 true, 则表示路由离开后页面 DOM 被销毁。

注意, 如果业务组件使用 React Hooks, 则需要注意 useEffect 模拟的 componentDidMount 在获取 props 时仅能获取到第一次渲染的 props (相关部分在 3.2.2 节曾有介绍)。在本例中, 在回调函数中获取路由匹配对象时, 建议从第一个参数 routeProps 中获取。

上述回调函数能兼容 Route 的 render、children 等不同渲染场景, 在使用上述方法时, 不用关心组件的生命周期, 以及 DOM 是否保存等, 如以下 A 组件示例:

```
export function A(props) {
  useEffect(function componentDidMount() {
    // 返回取消监听函数
```

```js
  /*
    didEnter 表示已经进入 Route 对应页面，在回调函数中传入了当前 routeProps 与
进入模式
    更新进入-->DOM 不销毁，重新进入-->初始挂载
  */
  const unlisternEnter = props.didEnter((routeProps, isUpdate) => {
    /*
      由于 React Hooks 的 Capture Value 特性，使用 props 将会只读到第一次渲染的 props，
可能造成问题
      这里需要使用 didEnter 回调函数中的 routeProps
      routeProps 在 6.3 节曾介绍过
     */
    console.log(
      routeProps.match.url + "进入了",
      isUpdate ? "更新进入" : "重新进入"
    );
  });
  // 返回取消监听函数
  /*
    onLeave 表示组件即将离开当前 Route 页面，isUnmount 表示组件在离开页面后是进
行销毁还是 CSS 隐藏
   */
  const unlisternLeave = props.onLeave((routeProps, isUnmount) => {
    console.log(
      `${routeProps.match.url}退出了`,
      isUnmount ? "组件即将销毁" : "组件隐藏"
    );
  });
  return function componentUnmount() {
    unlisternEnter && unlisternEnter();
    unlisternLeave && unlisternLeave();
  };
  // 在此处使用[]，仅 componentDidMount 触发
}, []);
return props.children;
}
```

9.4.2 实现路由生命周期高阶组件

为了实现上述能力，考虑使用高阶组件，在高阶组件中处理路由生命周期逻辑：

```
// 包装组件，为组件注入生命周期扩展
const Comp = withRouteLifeCycle(
  component
);
function App() {
  return (
    <BrowserRouter>
      <Route
        path="/list-C"
        // 支持 component 销毁组件的生命周期
        component={Comp}
      />
      <Route
        path="/list-C"
        children={props => {
          return (
            <div
              style={{
                display: props.match
                  ? "block"
                  : "none"
              }}
            >
              {/* 支持组件不销毁时的生命周期*/}
              <Comp {...props} />
            </div>
          );
        }}
      />
    </BrowserRouter>
  );
}
```

具体在实现时，先使用 React.memo（在 3.4 节曾有过介绍）记录原始组件信息。这一步的目的是当路由匹配命中时，才触发组件的 render，进而才触发子组件的 render。React.memo 与如下 shouldComponentUpdate 类似：

```
import * as React from 'react';
import { RouteChildrenProps } from 'react-router';
// 用于包裹业务组件，事实上当 nextProps.match 路由匹配命中时才需要渲染
export default class ComponentMatched extends React.Component<RouteChildrenProps> {
  shouldComponentUpdate(nextProps: RouteChildrenProps) {
    // 路由匹配命中时才需要执行 render
    // 路由未匹配命中时不需要执行 render
    return nextProps.match;
  }
  render() {
    return this.props.children;
  }
}
```

由于在路由匹配命中时才调用 render，从而保证了业务组件 render 中相关的路由操作不会报出错误（但通过记忆化后的组件需要传入 match 对象用于渲染判断）。业务组件永远在路由匹配命中时才进行渲染，如在使用 Route 的 children 渲染隐藏 DOM 模式时，由于 chilren 将无条件地被调用，会始终渲染组件。任何一次路由的变化，都会使得组件进入更新生命周期再触发 render，这时通过父组件的 shouldComponentUpdate，可阻止 render 的多次调用，也避免了如 props.match 为 null 时，props.match.params.id 的访问报错。

实现 withRouteLifeCycle 的主要思路是：withRouteLifeCycle 提供 didEnter、onLeave 注入组件。由高阶组件提供调用时机，子组件提供调用实现。在调用时机的选择上，通过 React.useContext 获取组件树中最近 Route 的匹配情况，并获得 Ref 记录的上一次 Route 的匹配情况，通过前后 Route 匹配结果的对比，可获知路由的变化，进而判断出是进入路由还是退出路由，并触发相关的回调函数：

```
import {
  __RouterContext as RouterContext,
  RouteComponentProps
} from "react-router";
import * as React from "react";
import { useContext, useEffect, useCallback, useLayoutEffect } from "react";
// 深度比较
import { isEqual } from "lodash";
interface onLeaveFuncType {
```

```
  (props?: RouteComponentProps, isUnmount?: boolean): void;
}
interface didEnterFuncType {
  (props?: RouteComponentProps, isUpdate?: boolean): void;
}
function noop() {}
// 高阶组件提升 cls 能力
export default function withRouteLifeCycle(cls) {
  // 记忆业务组件，仅在 match 有值时才渲染组件
  const Cls = React.memo(cls, (_, nextProps) => !nextProps.match);
  return props => {
    // 记录上一次命中情况
    const previousMatchRef = React.useRef(null);
    const routeContext = useContext(RouterContext);
    // 从 RouterContext 中获取本次命中情况
    const currentMatch = routeContext.match;
    const previousMatch = previousMatchRef.current;

    // 用于记录 didEnter、onLeave 回调的 refs
    const enter = React.useRef<didEnterFuncType>(noop);
    const leave = React.useRef<onLeaveFuncType>(noop);

    function didEnter(cb: didEnterFuncType) {
      // 保存组件的 didEnter 回调
      enter.current = cb;
      return () => {
        enter.current = noop;
      };
    }
    function onLeave(cb: onLeaveFuncType) {
      // 保存组件的 onLeave 回调
      leave.current = cb;
      return () => {
        leave.current = noop;
      };
    }
    // 缓存 onLeave、didEnter 不用每次都传入新函数到组件中
    const cachedLeave = useCallback(onLeave, [leave.current]);
```

```js
    const cachedEnter = useCallback(didEnter, [enter.current]);

useEffect(
  () => () => {
    leave.current(
      { ...routeContext, location: routeContext.history.location, match: previousMatchRef.current },
      // 当组件卸载时,设置 unMount 标记
      true
    );
    enter.current = noop;
    leave.current = noop;
  },
  []
);
/*
  使用 useLayoutEffect 是为了在浏览器渲染绘制前触发,可以在其中进行 DOM 操作,
  跟 onLeave 即将离开一致
*/
useLayoutEffect(function useLayoutEffectCallBack() {
  // 上一次渲染命中,本次未命中
  if (!currentMatch && previousMatch) {
    // 触发退出回调,组件没有卸载,当前未命中, currentMatch 为 null
    leave.current({ ...routeContext, match: previousMatch }, false);
  }
});
useEffect(() => {
  /*
    路由前后不相等(deep comparison)触发生命周期,此操作保证了仅在路由变化时触发回调
  */
  if (currentMatch && !isEqual(currentMatch, previousMatch)) {
    if (previousMatch) {
      /*
        使用 Promise.resolve 延迟执行 enter,晚于 React 的函数作用范围
        目的是使得任何 didEnter 回调执行都在 onLeave 回调之后
        保证执行顺序 onLeave---> didEnter
      */
      Promise.resolve().then(() => {
```

```
            // 上次组件渲染 Route 也命中, 标记为更新, 第二个参数为 true, 并传入当前
            // 的 Route 上下文到回调函数中
            enter.current(routeContext, true);
          });
        } else {
          // 使得任何 didEnter 的回调执行都在 onLeave 回调之后
          // 上次路由渲染未命中
          Promise.resolve().then(() => {
            // 进入 hook
            enter.current(routeContext, false);
          });
        }
      }
    });
    previousMatchRef.current = currentMatch;
    return <Cls {...props} onLeave={cachedLeave} didEnter={cachedEnter} />;
  };
}
```

上述 withRouteLifeCycle 的实现在三处使用了 React.useRef（React.useRef 曾在 3.2.4 节中有过介绍），目的是获得可持久保存的变量。其中，previousMatchRef 用以记录上一次渲染的路由匹配结果，以方便两次渲染间的路由匹配情况比对；另外两处创建的 Ref 用于保存子组件的回调函数，分别保存 didEnter 与 onLeave 的回调，以便高阶组件在适当时机被触发。

在 didEnter 生命周期逻辑处理上，选择在 effect 中进行，这符合 didEnter 已经进入路由的语义。为了触发 didEnter 回调，首先要判断当前路由上下文是否命中，只有路由命中才应触发 didEnter 回调；同时也需要判断前后匹配 currentMatch、previousMatch 是否相等，应该在 currentMatch 与 previousMatch 不相等时触发 didEnter 回调，这确保了 didEnter 回调在路由变化时触发。

同时考虑组件能匹配多个路径的情况，如：

```
<Route path="/user/:name" component={LifeCycleComponent}>
```

由于路径/user/a、/user/b 都会匹配路由规则，外部路由若从/user/a 跳转到/user/b，则 Route 都会匹配命中，且前后两次匹配结果都是命中的情况。此刻由于 previousMatch 也有值，可设置回调函数的第二个参数为 true，表明组件路由为更新进入。在上述情况下，/user/a 页面的 match.params.name 为 a，/user/b 页面的 match.params.name 为 b，通过注入的 routeProps.match.params 可获取到上述路由信息。

此外，在上述示例中还使用了 Promise.resolve 延迟 didEnter 回调的触发，这是为了保证在多个组件都使用此高阶函数包装时，生命周期的触发顺序：didEnter 回调的触发时机将晚于 onLeave 回调。

在 onLeave 生命周期逻辑处理上，选择在 2 个 effect 中进行。一个是 useLayoutEffect，当上次渲染路由匹配命中，而当前匹配未命中时，说明是从匹配路由离开，此刻可以调用通过 onLeave 所注册的回调函数，触发 onLeave 回调。选择在 useLayoutEffect 中触发回调的原因是，useLayoutEffect 将在页面绘制前触发，浏览器此时并未绘制，符合 onLeave 即将离开的语义（离开路由后，浏览器还未绘制）。由于离开路由后，当前匹配 currentMatch 为空，为了获得 onLeaveFuncType 函数中的参数，可以使用 previousMatch 作为 routeProps 的 match。由于 previousMatch 曾命中过，保留有命中信息，因此这保证了可从 routeProps.match 中获取曾命中过的 URL。同时由于此时组件并未销毁，设置第二个销毁标志参数为 false。

为了兼容路由离开时组件销毁场景，另外一个 effect 用来在组件销毁时调用，目的是获得组件的 componentWillUnmount 生命周期。由于路由离开，当前 currentMatch 为 null，为了使得业务组件获得曾经命中过的 match 信息，注入的第一个参数的 match 同样为 previousMatch（注意这里需要从 previousMatchRef.current 中获取）。同时由于此时组件即将销毁，设置第二个销毁标志参数为 true。

9.5 React Router 状态同步 Redux

9.5.1 接入 connected-react-router

在使用 React Router 时，多数通过使用由 Route 注入的 props.location 或者使用上下文等方式获取路由信息。

在 Redux 中，用于维护状态的数据存储于 store 中。若希望 React Router 结合 Redux，则需要将路由的状态同步到 store 中，组件可通过 mapStateToProps 映射路由信息，之后便可从 props 中获取。对于此种情况，社区有 connected-react-router 用于解决路由状态与 Redux 结合的问题，下面具体介绍其原理及使用。

对于 Redux 项目，接入 connected-react-router 以获得路由能力有三处配置，分别是：

（1）根 reducer 接入名为 router 的子 reducer 用于更新 store。

（2）store 接入 routerMiddleware 中间件用于特定 action 处理。

（3）使用 ConnectedRouter 代替 Router，扩展 React Router 中 Router 的能力。

对于第一点，在根 reducer 处，可从 connected-react-router 中引入 connectRouter，并可注入一个名为 router 的子 reducer，如：

```
import { connectRouter } from 'connected-react-router'
const createRootReducer = combineReducers({
  count: counterReducer,
  /*
    其余 reducer……
  */
  // router reducer 注意名称 router 不能更改
  router: connectRouter(history)
})
```

其中 history 为路由系统的 history；同时，在创建 store 时，配置 routerMiddleware。store 的结构如下：

```
import { applyMiddleware, createStore } from 'redux'
import createRootReducer from './reducers'
import { routerMiddleware } from 'connected-react-router'
const store = createStore(
  createRootReducer(),
  preloadedState,
  applyMiddleware(
    routerMiddleware(history),
  ),
)
```

从 connected-react-router 中引入 routerMiddleware，并传入路由系统的 history。在获得了 store 后，在路由外层引入由 connected-react-router 提供的 ConnectedRouter 组件，搭建出整个应用的框架：

```
<Provider store={store}>
  <ConnectedRouter history={history}>
    <Route
      exact
      path="/"
      component={Home}
    />
```

```
    <Route
      path="/hello"
      component={Hello}
    />
    {/*..........*/}
  </ConnectedRouter>
</Provider>
```

其中 history 为路由系统所期望的 history。对于使用 connected-react-router 搭建的 Redux 项目，React Router 的 Router 路由器并没有使用到，其原因为 ConnectedRouter 组件已经提供了，在之后的分析中会进行阐述。

在组件框架搭建完成后，路由信息便可从 store 中读取，如：

```
const A = ({ pathname, search, hash }) => (
  <div>
    <div>
      pathname: {pathname}
      search: {search}
      hash: {hash}
    </div>
  </div>
)
const mapStateToProps = state => ({
  // 把 Redux 中的 router 状态传递给组件 props
  pathname: state.router.location.pathname,
  search: state.router.location.search,
  hash: state.router.location.hash,
})
export default connect(mapStateToProps)(A)
```

组件 A 通过使用 connect 方法包裹后便可获得 Redux 中 store 的路由信息。

如果需要改变浏览器地址，则同样有 3 种方式，除了使用 react-router 原生支持的 Link 或 this.props.history.push(replace)，还可以从 connected-react-router 中引入对应的 action，如：

```
import {push, replace, go, goBack, goForward} from 'connected-react-router'
dispatch(push('/a'))
```

当发出 dispatch(push('/a')) 的 action 时，也会达到与 this.props.history.push('/a') 相同的效果。无论使用哪种跳转方式，最后都会发出与如下形式类似的 action：

```
const actionA = {
  type:'@@router/LOCATION_CHANGE',
  payload:{
      location:{
        pathname:'/a'
        ..........
      },
      action:'PUSH'
  }
}
```

同时,无论是以上哪种方式,在 Redux 的 store 中都能同步 type 为@@router/LOCATION_CHANGE 的 action 的变化,会将 payload 同步到 store 中,可从 store.getState 中获得最新的路由信息。

9.5.2　connected-react-router 原理分析

下面将分析上述 3 个步骤背后的原理。

1. connectRouter

connectRouter 的能力是提供与路由相关的 reducer,使得路由状态能同步到 Redux 的 store 中。之所以需要路由的 history 对象,是因为第一次渲染需要从 history 中获取当前 URL 的值。

```
const createRouterReducer = (history) => {
  const initialRouterState = fromJS({
    location: history.location,
    action: history.action,
  })
  return (state = initialRouterState, { type, payload } = {}) => {
    if (type === '@@router/LOCATION_CHANGE') {
      const { location, action } = payload
      return merge(state, { location: fromJS(location), action })
    }
    return state
  }
}
```

2. routerMiddleware

当想要产生路由变化时，可通过如下方式发出 action：

```
import {push, replace, go, goBack, goForward} from 'connected-react-router'
dispatch(push('/a'))
```

对应在业务代码中，当使用 dispatch 方法分发一个 connected-react-router 中的 action 时，会产生 type 为 @@router/CALL_HISTORY_METHOD 的 action。

中间件会拦截这个 type，这里的中间件即 routerMiddleware。

```
const routerMiddleware = (history) => store => next => action => {
  if (action.type !== '@@router/CALL_HISTORY_METHOD') {
    return next(action)
  }
  const { payload: { method, args } } = action
  history[method](...args)
}
```

routerMiddleware 的主要功能是拦截 type 为 @@router/CALL_HISTORY_METHOD 的 action，并执行对应的 history 方法，进行状态更新。

3. ConnectedRouter

由于浏览器路径的变化可以通过 history.listen 函数监听，因此在 ConnectedRouter 内部可以利用此接口监听路径的变化，发出一个自定义的 action。再利用之前提到的名为 router 的 reducer 更新 store 的状态，便可同步路由信息到 Redux 的 store 中。

```
class ConnectedRouter extends PureComponent {
  constructor(props) {
    super(props)
    const { history } = props
    const handleLocationChange = (location, action, isFirstRendering = false) => {
      /*
          在地址发生变化时，发出对应的 action，省略具体实现
      */
    }
    this.unlisten = history.listen(handleLocationChange)
    handleLocationChange(history.location, history.action, true)
```

```
    }
    componentWillUnmount() {
      this.unlisten()
    }
    render() {
      const { history, children } = this.props
      return (
        <Router history={history}>
          { children }
        </Router>
      )
    }
}
```

由以上分析可知，ConnectedRouter 扩展了 react-router 包中 Router 的能力。

9.6 React Router 状态同步 Mobx

在知道了如何将路由状态同步到 Redux 后，将状态同步到 Mobx 就变得非常容易。社区也有类似的第三方库，如 mobx-react-router。实现思路为：监听 history 的变化，并同步状态到 store 中。这里以 mobx-react-router 为例。由于可以在外部添加监听事件，且状态的同步与组件或 UI 没有关联，因此可以为 history 封装一个修饰函数，用于给 history 添加功能：

```
export const decoratedHistory = (history, store) => {
  const handleLocationChange = (location) => {
    //更新 store
    store.updateLocation(location);
  };
  //当路径变化时更新 store 的值
  history.listen(handleLocationChange);
  handleLocationChange(history.location);
  return history;
};
```

使用 Router 的 history 监听路径的变化，当路径变化时，更新 store 的值。

```
// mobx store
import { observable, action } from 'mobx';
```

```js
export class RouterStore {
  @observable location = null;
  @action
  updateLocation(newLocation) {
    this.location = newLocation;
  }
};
```

由于 updateLocation 为一个 action 函数，其改变了 observable 的 location 变量，相应地观察 location 的位置可得到通知。在使用时，仅需要把 Router 的 history 用 decoratedHistory 函数包裹一下。

```js
export const routerStore = new RouterStore();
const history = createBrowserHistory();
<Router history={decoratedHistory(history, routerStore)} />
  /* 便可通过 routerStore 获得对应的 location 信息
  routerStore.location.pathname
  routerStore.location.search
  routerStore.location.hash
*/
```

9.7 路由与组件的结合实战

9.7.1 路由结合 Tabs 组件

以 antd 的 Tabs 组件为例，提供 withRouterTabs 高阶组件，使得 Tabs 可与 URL 的 query 进行联动：

```js
import { Tabs } from "antd";
const RouterTabs = withRouterTabs(Tabs);
<RouterTabs queryKey="tab" defaultActiveKey="1">
    <TabPane tab="Tab 1" key="1">
      Content of Tab Pane 1
    </TabPane>
    <TabPane tab="Tab 2" key="2">
      Content of Tab Pane 2
    </TabPane>
```

```
    <TabPane tab="Tab 3" key="3">
      Content of Tab Pane 3
    </TabPane>
</RouterTabs>
```

在设计上，使用 TabPane 的 key 值作为 URL 中 query 的值，如?tab=3；也可使用属性 props. queryKey 作为 URL 中 query 的 key 值。

这样在使用时，当单击某个 Tab 时，URL 的 query 也将得到更新，如图 9-1 所示。

图 9-1 Tabs 组件与路由

使用 React Hooks 开发高阶组件，先进行一些准备工作，声明自定义 Hook——useQuery 用于获得浏览器的 query：

```
export const useQuery = () => {
  const location = React.useContext(__RouterContext).location;
  const { search } = location;
  const query = useMemo(() => queryString.parse(search), [search]);
  return query;
};
```

同时声明 useUpdateQuery 自定义 Hook 用于获得更新 query 的函数：

```
export const useUpdateQuery = options => {
  // 获取 history
  const { history } = React.useContext(__RouterContext);
  // 使用 useQuery 获得 query
  const query = useQuery();
  const { replace } = options;
  const updateQuery = useCallback(
    updateQuery => {
      // 更新 query
      const newQuery = { ...query, ...updateQuery };
      const newSearch = queryString.stringify(newQuery);
```

```
    // 只使用 history 修改 search
      if (replace) {
        history.replace({ search: newSearch });
      } else {
        history.push({ search: newSearch });
      }
    },
    [history, query, replace]
  );
  return updateQuery;
};
```

其 options 可设置为 replace 为 true，则替换 URL 的 query 的行为使用 history.replace，默认为 history.push。在上例自定义 Hook 中，使用 useQuery 获取现有 query，useCallback 用于保存函数副本，避免每次都生成函数，并确保依赖为[history, query, replace]，以便依赖变化函数得以更新。在函数实现中，使用 queryString 将传入 query 与现有 query 合并，作为新的 history 方法的入参。

在准备好自定义 Hook 之后，可进行高阶组件的编写：

```
function withRouterTabs(Tabs) {
  return props => {
    const originOnChange = props.onChange;
    const updateUrlQuery = useUpdateQuery({ replace: true });
    const search = useQuery();
    // 默认的 key 为_tabKey
    const queryKey = props.queryKey || "_tabKey";
    const tabkey = search && search[queryKey];
    // 声明新的 onChange 的代理函数传入 Tabs
    const onChange = useCallback(
      (...args) => {
        originOnChange && originOnChange(...args);
        const key = args[0];
        args && key && updateUrlQuery({ [queryKey]: key });
      },
      [originOnChange, updateUrlQuery, queryKey]
    );
    // 在初始挂载时，人工触发一次变更
    useEffect(() => {
```

```
      !tabkey && onChange(props.defaultActiveKey);
    }, []);
    return (
      <Tabs
        {...props}
        onChange={onChange}
        activeKey={tabkey || props.defaultActiveKey || props.activeKey}
      >
        {props.children}
      </Tabs>
    );
  };
}
```

高阶组件 withRouterTabs 接收原有 Tabs 组件, 并返回新函数组件, 用于扩展原 Tabs 的能力。使得 Tabs 组件与路由结合的关键在于, 为 Tabs 组件引入内部的属性, 并覆盖原属性:

```
    {/* 真正的 Tabs 组件 */}
<Tabs
    {...props}
    onChange={onChange}
    activeKey={tabkey || props.defaultActiveKey || props.activeKey}
>
    {props.children}
</Tabs>
```

声明了 onChange, 将 tabkey 传入 Tabs 组件中。activeKey 由组件内部控制, 根据 Tabs 组件的 onChange 第一个参数是 TabPane 的 key 值的特点, 获取到 key 值后, 便可调用 updateUrlQuery 更新 query:

```
    const onChange = useCallback(
      (...args) => {
        originOnChange && originOnChange(...args);
        const key = args[0];
        args && key && updateUrlQuery({ [queryKey]: key });
      },
      [originOnChange, updateUrlQuery, queryKey]
    );
```

如果组件传入过 onChange, 则也将进行代理触发。使用 useCallback 是为了记录 onChange

的副本，避免多次创建 onChange 函数。

与此同时，还需要解决数据到组件的同步问题，包括第一次初始渲染或者使用 history 方法引起 query 改变，当读取到 query 时，Tabs 都会进行渲染，通过 useQuery 获得 query，设置到 activeKey 中：

```
const search = useQuery();
  const queryKey = props.indentify || "_tabKey";
  const tabkey = search && search[queryKey];
/*
    使用 tabkey 作为 Tabs 的激活 key
*/
```

还有一种情况是，当 URL 没有 query 时，由于读取不到激活的 key 值，这时如果不进行处理，则将不会激活任何 TabPane。这可以在挂载成功时，人工触发一次 defaultActiveKey 的变化：

```
useEffect(() => {
    !tabkey && onChange(props.defaultActiveKey);
  }, []);
```

如果传入了 defaultActiveKey 属性，则会触发一次 onChange 更新操作，会调用 updateUrlQuery 更新 query，URL 将会发生变化。

9.7.2 路由结合 Modal 组件

使用路由控制弹窗 Modal 的好处是，弹窗状态不用设置在内存中，而是由外部控制，如在页面跳转过程中需要保持弹窗，仅需要控制 URL 即可。

在实现上，使用 Route 的 children 属性渲染 Modal，Route 的作用为匹配器，通过上下文中的 match 获取到组件树中最近一级的 Route 的命中情况，并拼接上自身的 URL 作为 Route 的 path。如当前 URL 为/foo，Modal 的 path 为/modal，则 Modal 的路径为/foo/modal。当命中的状态为 isExact，表示路由绝对匹配时，可控制 Modal 进行显示；反之，未命中路由则隐藏 Modal。

```
function showModal(url, routeContext) {
  const modalUrl = `${routeContext.match.url}${url}`;
  routeContext.history.push(modalUrl);
}
function RouteModal(modalProps) {
  const { history, match } = useContext(RouterContext);
  <Route
```

```
      path={match && `${match.path}${modalProps.path}`}
      children={routeProps => {
        <Modal
          {...modalProps}
          show={routeProps.match && routeProps.match.isExact}
          onCancel={history.goBack}
        />;
      }}
    />;
}
```

使用 history.push 方法可改变 Modal 的展示状态，仅需在当前页面定义的 Route 中获取到匹配的 URL，并拼接 Modal 的后缀地址，如当前页面定义的 Route 为/foo, showModal("/modal", …) 则会将当前 URL 地址更改为/foo/modal，以控制 Modal 的展示状态。

9.7.3　路由结合 BreadCrumb 组件

BreadCrumb 组件也称面包屑组件，通常在面包屑导航中使用。面包屑导航是用户界面中的一种辅助导航。通常面包屑导航在页面顶部水平出现，一般会位于标题或页头的下方。它们提供给用户返回之前任何一个页面的链接（这些链接也是能到达当前页面的路径）。面包屑导航提供给用户回溯到网站首页或入口页面的一条路径，通常以大于号（>）出现。它们绝大部分看起来就像这样：首页>分类页>次级分类页。本节将介绍面包屑导航的高阶组件 withBreadcrumbs。

面包屑组件通常用于标识当前的路径与上级各路径。本节设计纯逻辑的高阶组件，用以提供路径信息，UI 部分可使用高阶组件所提供的信息进行渲染。在实现上，通过引入 withBreadcrumbs 高阶函数，支持装饰各类 UI 组件 BreadCrumbs：

```
import { Link } from "react-router-dom";
import withBreadcrumbs from "./breadcrumb";
// Breadcrumb 的签名，有 url 与 name 属性
export interface Breadcrumb {
  url: string;
  name: string;
}
// BreadCrumbs 组件示例
// 面包屑纯 UI 组件，可使用 breadcrumbs 属性渲染面包屑导航项
function BreadCrumbs(props: { breadcrumbs: Breadcrumb[] }) {
```

```
  // 通过 withBreadcrumbs 注入 breadcrumbs
  return props.breadcrumbs.map(breadcrumb => {
    // 可渲染每个导航 Link
    return <Link to={breadcrumb.url}>{breadcrumb.name}</Link>;
  });
}
// 面包屑 UI 组件 BreadCrumbs 经 Hoc 包装后
const BreadCrumbsHoc = withBreadcrumbs()(BreadCrumbs);
function A() {
  return <BreadCrumbsHoc />;
}
```

引入 withBreadcrumbs 高阶函数，通过 withBreadcrumbs 返回的高阶组件将获得 props.breadcrumbs，可用其渲染面包屑导航。

为了实现 withBreadcrumbs，使用 withRouter 或 Hook 获得当前的 location，或者可理解为监听 location 的变化，从中能获得最新的当前 location 信息。据此可将 location 按照 "/" 分成多个 segment，如将/1/2/3 分成 1,2,3 之后可通过 reduce 或其他方式拼接得到导航链接/1、/1/2、/1/2/3，在将导航链接收集到数组并注入 UI 组件中后，组件可从 props 中获取到一个如[/1,/1/2,/1/2/3]的数组，进而可使用该数组渲染导航项。withBreadcrumbs 的实现如下：

```
import { useLocation, matchPath } from "react-router";
export interface Breadcrumb {
  url: string;
  name: string;
}
interface config {
  // 配置的匹配路径
  path: string;
  // 该配置是否希望排除
  exclude?: boolean;
  // 配置的注入名
  name?: string;
}
export function stripTrailingSlash(path) {
  return path.charAt(path.length - 1) === "/" ? path.slice(0, -1) : path;
}
function getSegments(pathname: string, configs?: config[]): Breadcrumb[] {
  // 去掉结尾的 "/"，如果为根路径，则会得到空字符串
```

```
    const currentPathname = stripTrailingSlash(pathname);
    const breadcrumbs: Breadcrumb[] = [];
    currentPathname
      // 按 "/" 分割各个 segment
      .split("/")  // 如果字符串为空字符串，则会得到空数组
      .reduce((previous, current) => {
        // 若为空字符串，则在此处提供根路径 "/"，currentPath 在每次迭代中，顺序为 /1、
        // /1/2、/1/2/3
        const currentPath = current ? `${previous || ""}/${current}` : "/";
        let finedConfig;
        if (configs && configs.length) {
          // 寻找能匹配命中当前路径的配置
          finedConfig = configs.find(config => {
            return matchPath(currentPath, {
              path: config.path,
              exact: true
            });
          });
        }
        // 不用再次提供根路径
        const pathIgnoreRoot = currentPath === "/" ? "" : currentPath;
        // 若配置为 "排除"，则不加入 breadcrumbs 数组
        if (finedConfig && finedConfig.exclude) {
          return pathIgnoreRoot;
        }
        // 高阶组件提供面包屑导航的数据支持
        breadcrumbs.push({
          url: currentPath,
          // 配置提供了命名
          name: (finedConfig && finedConfig.name) || current || "/"
        });
        return pathIgnoreRoot;
      }, null);
    return breadcrumbs;
}
export default (configs?: config[]) => Component => {
  return props => {
    const currentLocation = useLocation();
```

```
    // 根据当前的 location 获取各个 segment 作为 breadcrumbs 数组
    const breadcrumbs = getSegments(currentLocation.pathname, configs);
    return <Component {...props} breadcrumbs={breadcrumbs} />;
  };
};
```

上例在实现上，可通过调用 getSegments 得到的面包屑数组，并注入组件中，获得如[/,/a,/a/b]这样的数组。在一些情况下，渲染导航项希望有自定义的命名，如 "/" 路径在无额外配置的情况下注入的导航名称同样为 "/"；或者希望能绕过一些导航项，如在路径/user/1 下提供的面包屑数据为[/,/user,/user/1]，但若业务中/user 路径无意义，则 UI 组件不希望关心/user 路径，仅希望获得[/,/user/1]这样的路径。这两种情况都可以通过高阶函数的配置实现：

```
interface config {
  path: string;
  exclude?: boolean;
  name?: string;
}
const BreadCrumbsHoc = withBreadcrumbs([
  { path: "/", name: "首页" },
  {
    path: "/user",
    exclude: true
  }
])(BreadCrumbs);
```

其中对路径 "/" 设置了注入的导航名，当命中 "/" 时，导航项将提供 "首页" 名称进行渲染，且对路径/user 进行 "排除"，注入组件的导航列表中将无/user 导航项，上例同样实现了此能力。

9.8 为 history 方法引入前置中间件

9.8.1 Redux 中间件

在 Redux 中有中间件的概念，Redux 的每个 action 会依次通过中间件。在 Redux 应用中可提供多个中间件，并可在各中间件中做一些处理，如 redux-thunk 和 redux-promise。借鉴 Redux，也可为 history 等方法引入中间件机制。

如图 9-2 所示，history.push 等导航方法将通过各个中间件，中间件通过调用 await next()进入下一个中间件，并在所有中间件都通过后才执行原始 history 方法。由于对 history 的 push、replace、go、goBack 和 goForward 方法引入了中间件，因此每个中间件可获得对应方法的信息，如 path 和 state 等，可在中间件中做相关的逻辑处理，也可以选择不进入下一个中间件，这样将不会执行导航函数。

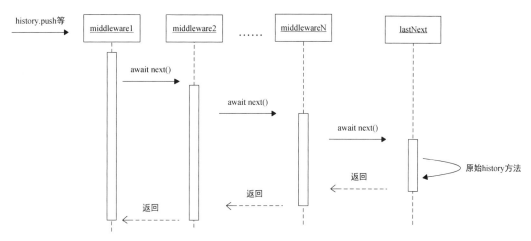

图 9-2 history 各中间件接管导航的执行

9.8.2 中间件定义

在使用上，通过 applyMiddleware 为 history 引入各中间件：

```
applyMiddleware(history, [middware1, middware2, middware3]);
```

为 history 中间件引入异步机制，可在中间件中执行异步的方法。中间件的写法如下：

```
// 调用 next 表明进入下一个中间件，与 Redux 类似
export async function middware1(next, state) {
  console.log("第一个中间件");
// 如果希望中间件拦截 history 等方法，则不用调用 next
  await next();
}
```

其中，参数 next 代表执行下一个中间件，其签名如下：

```
next: (redirect?: Location & {push?:boolean} | string) => void,
```

next 的第一个参数代表需要重定向的路径，可以为空，接收 Location 对象和字符串类型，默认的 redirect 方式是 replace，可以设置 push 为 true，使用 history.push 进行重定向。与 Redux 类似，如果直接调用 await next()，则将会进入下一个中间件。

中间件的第二个参数为 state，其包含调用 history 方法时的众多信息，签名如下：

```
interface MiddlewareState {
  history?: History;
  method?: "push" | "replace" | "go" | "goBack" | "goForward";
  from?: Location;
  to?: Location | string | number;
  originParams?: any[];
}
```

其中，Location 为 history 中的 Location：

```
export interface Location<S = LocationState> {
    pathname: Pathname;
    search: Search;
    state: S;
    hash: Hash;
    key?: LocationKey;
}
```

中间件函数的参数 state 对象包含了 history 对象，可从中获取路径未改变前 history.length、history.state 等信息。也可从中获取 method 方法，得知调用导航改变的具体方法名，并可通过 from 与 to 获得跳转的起始与目标地址。由于调用 history.go 等方法时无法获知目标 URL，所以 to 的类型也可为 number。最后，还可从 state 中获取原始传递给 history 方法的参数。如果调用 history.push(a,b,c,d)，则这里 originParams 将会以数组[a,b,c,d]保留。

由于每个 next 函数都为异步函数，所以调用 next 的中间件也同样需要为 async 异步函数，如：

```
// 不关心 state
export async function middware2(next) {
  console.log("第二个中间件");
  // 支持异步
  await new Promise(resolve => {
    setTimeout(() => {
      resolve();
    }, 600);
  });
```

```
    await next();
  }
```

上例中间件不使用第二个参数 state，可仅保留参数列表中的 next 函数。

如果希望中间对某个导航路径进行拦截重定向，则可在 next 函数中传入重定向字符串或 Location 对象：

```
// 如果未登录，且导航不是登录页，则重新导航到登录页
export async function middware3(next, state) {
  console.log("第三个中间件");
  const { to } = state;
  if (to !== "/login" && !globalLogin) {
    console.log("未登录，将导航到登录页");
    // 重新导航到登录页
    await next("/login");
  } else {
    await next();
  }
}
```

在上例中，当检测到导航路径不是/login 登录页，且全局登录状态为 false 时，调用 await next("/login")重新导航到登录页。这样将重新执行 history.replace("/login")，原有的导航请求将被忽略。

9.8.3　实现 history 中间件

实现 history 方法的前置中间件，可对 history 各方法进行包装重写：

```
import { History, Location, createPath } from "history";
import * as _ from "lodash";
interface MiddlewareState {
  history?: History;
  method?: "push" | "replace" | "go" | "goBack" | "goForward";
  from?: Location;
  to?: Location | string | number;
  originParams?: any[];
}

export interface redirectLocationDescriptorObject extends Location {
```

```
    push?: boolean;
}
type Middleware = (
    next: (redirect?: Location & { push?: boolean } | string) => void,
    state?: MiddlewareState
) => void;

// 提供重定向
async function redirectTo(
    history: History,
    redirect: redirectLocationDescriptorObject | string
) {
    let historyMehod = history.replace;
    let path: string;
    let state;
    if (typeof redirect !== "string") {
        state = redirect.state;
        path = createPath(redirect);
        if (redirect.push) {
            historyMehod = history.push;
        }
    } else {
        path = redirect;
    }
    return await historyMehod(path, state);
}
export default function applyMiddleware(
    history: History,
    middlewares: Middleware[]
) {
    const methods = ["push", "replace", "go", "goBack", "goForward"];
    // 用以保存原始的导航方法
    const originHistoryMethods = {};
    methods.forEach(method => {
        originHistoryMethods[method] = history[method];
    });
    let loading = false;
```

```
  // 对 methods 每一个方法都进行重写
  methods.forEach(method => {
// 重写 history 方法
    history[method] = async function(...args: any[]) {
      if (loading) {
        // 如果异步调用之后还未返回, 则不进行导航操作
        return;
      }
      // 最末尾的一个 next 函数
      let next = async (
        redirect?: redirectLocationDescriptorObject | string
      ) => {
        // 最后一个中间件
        if (redirect) {
          loading = false;
          await redirectTo(originHistoryMethods, redirect);
          // 返回, 不执行真正的 history 方法
          return;
        };
loading = false;
        // 将真正的 history 等方法放在最后一个中间件内执行, 原有参数透传到方法参数列表中
        await originHistoryMethods[method](...args);
      };
      // 当前的 history.location 即 from 的值
      const from = history.location;
      const to = args[0];
      // 从右到左遍历集合中的每一个元素, 从右到左依次更新 next 函数, 每个函数的 next 由
      // 上一个 middleware 进行 bind 调用获得
      // 除了 next, 依次将 history、当前方法 method、当前起始 from、目标路径 to, 以
      // 及整个调用方法的参数数组封装成 state 对象传递给函数
      _.forEachRight(middlewares, middleware => {
        // 使用 bind 将 next 函数设置为中间件的第一个参数
        const fn = middleware.bind(undefined, next, {
          history,
          method,
          from,
          to,
```

```
          args
        });
        // 为 next 函数包裹外层参数
        next = async (redirect?: redirectLocationDescriptorObject | string) => {
          // 根据参数进行重定向跳转等
          if (redirect) {
            loading = false;
            // 重新开始
            await redirectTo(originHistoryMethods, redirect);
            // 返回，不执行真正的 next 函数
            return;
          }
          loading = true;
          // 真正的 next 函数
          await fn();
          loading = false;
        };
      });
      await next();
    };
  });
}
```

初始将原始 history 各方法的引用进行保留，并重写 history 各方法。设置一个内部的 next 方法，在调用链最末尾调用，并将真正产生导航的方法置于最末尾的 next 函数内，这里为：

```
    let next = async (
      redirect?: redirectLocationDescriptorObject | string
    ) => {
      console.log("最后一个中间件");
      if (redirect) {
        loading = false;
        await redirectTo(history, redirect);
        // 返回，不执行真正的 history 方法
        return;
      }
loading = false;
      // 将真正的 history 等方法放在最后一个中间件内执行，原有参数透传到方法参数
      // 列表中
```

```
      await originHistoryMethods[method](...args);
    };
```

如果某一个中间件不执行 next 函数，那么就到不了最后这个 next 函数内执行原始的 originHistoryMethods。

类似 Redux、Koa，包裹中间件的思路是不断地更新 next 函数，每一次调用 next 函数即调用下一个中间件。

在实现上，从右到左遍历中间件数组中的每一个中间件，从右到左依次更新 next 函数，每个函数的 next 由上一个 middleware 进行 bind 调用获得，最后一个中间件的 next 由内部提供。除了 next，依次将 history、当前方法 method、当前起始 from、目标路径 to，以及整个调用方法的参数数组封装成 state 对象传递给函数：

```
    _.forEachRight(middlewares, middleware=> {
      // 使用 bind 得到无参数的 next 函数
      const fn = middleware.bind(undefined, next, {
        history,
        method,
        from,
        to,
        args
      });
      // 为 next 函数包裹外层参数，更新 next 引用
      next = async (redirect?: redirectLocationDescriptorObject | string) => {
        // 根据参数进行重定向跳转等
        if (redirect) {
          loading = false;
          // 重新开始
          await redirectTo(history, redirect);
          // 返回，不执行真正的 next 函数
          return;
        }
        // 真正的 next 函数
        await fn();
      };
    });
    loading = true;
```

```
        // 最后一个 next 函数，执行第一个中间件
        await next();
```

在上例中，为中间件进行 bind 调用得到了 fn 函数，这里的 fn 为中间件的别名，调用 fn 即调用中间件。此时更新 next 引用并对 fn 进行包装，在这个新的 next 函数中可接收第一个参数判断是否需要重新跳转，即中间件中的 next("/login") 用法。如果中间件调用方没有传入第一个参数，则会执行 await fn()，调用下一个中间件。

在从右到左依次更新完 next 函数后，最终得到的 next 函数为第一个中间件的 bind 调用返回，调用 await next() 将触发第一个中间件的执行。这样调用改造后的如 history.push 方法将触发第一个 next 函数，而后需要在每个中间件内部调用 await next()，直到最末尾的 next 执行真正的导航函数。

9.9 组件路由化

9.9.1 为组件加入 path 属性

对于一些轻量应用来说，引入 Route 用于申明路径有些累赘，事实上路由 Route 都是匹配路径进行组件渲染的，如果能直接在组件上声明路由路径，则代码会更加清晰简洁：

```
<App>
<User path="/user"/>
<Setting path="/setting"/>
</App>
```

User 的子组件也可以声明 path 路径，如：

```
<App>
<User path="/user">
<UserInfo path="/info"/>
</User>
<Setting path="/setting"/>
</App>
```

这样，当路径为 /user 时，会渲染 User 组件；当路径为 /user/info 时，会渲染 User 和 UserInfo 组件。

为了实现此种写法，需要由某个外层组件改写相关的业务组件，其本质是为每个有 path 属性的组件外层套上 Route 组件，得到 Route 的能力，如图 9-3 所示。

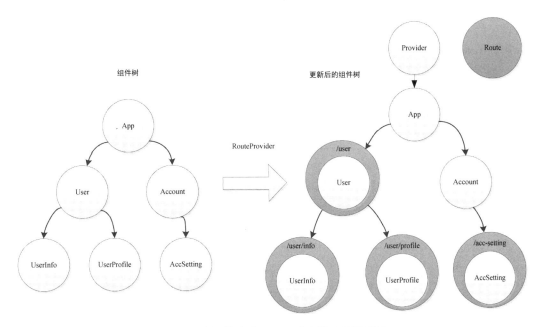

图 9-3　为组件套上 Route 获得路由匹配能力

9.9.2　为组件赋予路由

在实现手段上，为了使得组件获得路由化能力，需要为业务提供外层 Provider 组件。Provider 组件的功能是改写其可触达的各个子组件，使得各个子组件及后代组件获得路由能力：

```
interface Context {
  path?: string;
}
// 路径 Context
const RouteProviderContext = React.createContext<Context>({ path: "/" });
// 高层父组件，用于提供包装支持
export default function RouteProvider(props) {
  const reactRouterContext = React.useContext(__RouterContext);
  // 通过 history 的有无，判断是否应该实例化顶层 Router
  const history = reactRouterContext && reactRouterContext.history;
```

```
const ReactRouter = history
  ? (props: { children: React.ReactElement }) => {
      return props.children;
    }
  : Router;
// 使用 WrapComponentEachChild 提升子组件
return (
  <ReactRouter history={props.history}>
    <WrapComponentEachChild>{props.children}</WrapComponentEachChild>
  </ReactRouter>
);
}
```

通过 RouteProvider，为组件提供路由化支持：

```
<App>
<RouteProvider history={history}>
<User path="/user"/>
<Setting path="/setting"/>
</RouteProvider>
</App>
```

与 Router 一样，RouteProvider 接收 history 对象，且在内部实例化 Router，并且使用 WrapComponentEachChild 进行子组件的遍历包装。如果上下文中已经有 Router 的 history，则说明 Router 已经实例化，这时不做任何处理，返回 props.children。之所以这样做，原因在后面的父子组件示例中会有介绍。

为了实现对每个组件包裹 Route，需要深度遍历组件树，并改写原组件。为了达到这个目的，通常的做法是，遍历某个组件下的 children，依次为每个符合条件的 child 元素包裹上外层父组件 Route，并递归此过程。在实现上，可以提供一个外层组件 WrapComponentEachChild 作为 props.children 的父组件：

```
import * as React from "react";
import { __RouterContext, Router, Route } from "react-router";
import { omit } from "lodash";

export function stripTrailingSlash(path) {
  if (!path) return "";
  return path.charAt(path.length - 1) === "/" ? path.slice(0, -1) : path;
}
```

```js
export function addLeadingSlash(path) {
  if (!path) return "";
  return path.charAt(0) === "/" ? path : "/" + path;
}
export function isCompositeTypeElement(element) {
  return React.isValidElement(element) && typeof element.type === "function";
}
export function WrapComponentEachChild(props) {
  // 使用 map 方法保证 child 都有 key 值
  const wrappedChildren = React.Children.map(props.children, childElement => {
    return <WrapComponent>{childElement}</WrapComponent>;
  });
  return <>{wrappedChildren}</>;
}
// 所有 element 都应该包裹 WrapComponent
function WrapComponent(props) {
  const context = React.useContext(RouteProviderContext);
  const element = props.children;
  if (typeof element !== "object") {
    // 如果是基本元素、数字、字符串,则直接返回
    return element;
  }
  const path = element && element.props && element.props.path;
  const remainDom = element && element.props && element.props.remainDom;
  // 拼接上下文中的 path 与本级 path,构成当前路径
  const currentPath = `${stripTrailingSlash(
    addLeadingSlash(context.path)
  )}${addLeadingSlash(path)}`;
  // 获得 element 的 children
  const childElements = element && element.props && element.props.children;
  // 在使用 Route 提供路由支持的同时,也使用 WrapComponentEachChild 提升其各子组件
  const enhancedElement = React.cloneElement(
    element,
    {},
    childElements ? (
      // 使用 WrapComponentEachChild 提升 element 的 children
      <WrapComponentEachChild>{childElements}</WrapComponentEachChild>
    ) : (
      undefined
    )
```

```
  );
  /*
    通过判断是否有 path 属性且 element 是否是组件(类组件、函数组件)，来选择是否应该
在组件外包裹 Route
  */
  return path && isCompositeTypeElement(element) ? (
    // 为后代组件提供新的路径
    <RouteProviderContext.Provider value={{ path: currentPath }}>
      <Route
        // 还可接收 exact、strict 等参数
        {...omit(element.props, "component", "render")}
        path={currentPath}
        children={routeProps => {
          const routeEnhancedElement = React.cloneElement(
            enhancedElement,
            routeProps
          );
          /*
            业务组件除了 path，还可以传入 remainDom，这时组件无条件渲染
            同时业务组件接收到 routeProps，可自行判断渲染
          */
          if (remainDom) {
            return routeEnhancedElement;
          } else {
            return routeProps.match ? routeEnhancedElement : null;
          }
        }}
      />
    </RouteProviderContext.Provider>
  ) : (
    // 返回提升过的元素，使得递归子元素
    enhancedElement
  );
}
```

WrapComponentEachChild 为每个子元素套上 WrapComponent，以提供递归与路由支持。注意，上例使用 map 方法，目的是使无 key 的元素 child 产生新的 key。

WrapComponent 父组件在为元素提供 Route 能力的同时，对于每个被 WrapComponent 包裹

的元素，可以通过 React.cloneElement 更新其 children。这里对元素原有的 children 再次使用 WrapComponentEachChild 进行包裹，使得该元素下每个子元素外层同样套上 WrapComponent。通过这样递归的嵌套过程，从而使得所有元素外层都套上了 WrapComponent。

在 WrapComponent 中，props.children 即真实的业务组件，可从中获取到业务组件中的所有 props 信息。基于此，可以通过判断有无 path 属性，来选择是否嵌套外层的父组件 Route。如果拥有 path 属性，则可以通过 React.useContext(RouteProviderContext)获得最近一级的父路径信息 context.path，并拼接当前 path 路径，得到 Route 的最终渲染路径：

```
const currentPath = `${stripTrailingSlash(
  addLeadingSlash(context.path)
)}${addLeadingSlash(path)}`;
```

在 WrapComponent 中，通过 RouteProviderContext.Provider 更新当前路径，使得后代组件在构造路径时可进行路径拼接，形成嵌套路径。

这里在 WrapComponent 中，判断了每个 child 元素为复合元素（复合元素是 element 的 type 为函数的元素，可以是类组件或函数组件元素）。如果为复合元素且拥有属性 path，则可在原元素外层套上路由父组件进行支持。

注意，如果组件没有 children 元素，如 User 组件没有子组件：

```
<App>
<RouteProvider history={history}>
  {/* User 没有 children 元素 */}
<User path="/user"/>
<Setting path="/setting"/>
</RouteProvider>
</App>
```

而 User 的实现为：

```
function User(props){
  return <>
    {/* …… */}
      <UserInfo path="/info">
    {/* …… */}
  </>
}
```

此时，为 UserInfo 组件声明的路径/info 将不会生效。原因在于使用 RouteProvider 遍历组件并包裹 Route 时，仅对当前组件树生效。上例中当前组件树仅有 User 与 Setting 两个组件，因此在遍历时也只能遍历到这两个组件。事实上，User 并无 children，其 props.children 为空。由于无法遍历到 User 内部的 UserInfo，因此其/info 路径自然也不会生效。解决此问题需要在 User 内部使用 RouteProvider 提供包装支持：

```
function User(props){
  return <RouteProvider>
    {/* ................ */}
      <UserInfo path="/info">
    {/* ................ */}
    </RouteProvider>
}
```

由于上下文中已经存在 history，因此此时的 RouteProvider 等同于 WrapComponentEachChild。通过 WrapComponentEachChild 遍历每一个子组件并包裹 Route，从而使组件得到路由化支持。

本例也可以使用 React.useContext(__RouterContext).match 获得 match.path 作为父级的路径 path，这样的结果是子路由依赖于父路由的渲染结果。本例中使用了额外的上下文提供路径，这样做的好处是每个 Route 都是独立解耦的，父 Route 的命中结果不会影响子 Route。

9.10 路由与页签机制

9.10.1 页签介绍

页签（或者称为标签）在现代浏览器中被广泛使用，其形态通常如图 9-4 所示。

图 9-4 浏览器页签

在浏览器中，原生页签可以开启多个，即便域名与路径都一致，也可以存在多个页签。在一个页签下，可以容纳一个域名下的多个单页或多页页面，也可以容纳多个域名的多个页面。

在单页面应用 SaaS 系统中，可引入页签机制，模拟浏览器的页签效果。在设计时，每个业务多是处于一个页签下，每个页签通常容纳一个具体业务的全流程，如列表页、新增/编辑页、详情页等，都处于一个页签之下。使用页签的好处是用户可在不同业务间快速切换，以获得更好的应用内导航能力。

应用内页签不同于浏览器页签，其存在诸多限制，不能容纳多个域名（在不使用 iframe 方案的情况下），通常也不建议在页签内刷新跳转页面，页签数量也不如浏览器页签可无限添加（在内存允许的情况下）。其限制主要有：
- 页签内的页面不能跨域名（在无 iframe 的情况下）。
- 页签需要标志进行标识。
- 页签间的标志不能重复。
- 页签数量是固定的，有一定上限。

同一级别页签具有同一时刻仅有一个处于激活状态的特点，这与浏览器中同一时刻仅有一个 URL 的特点相吻合。通常，页签以 URL 中某个部分的值作为标志，与 URL 的某个部分形成一一对应的关系。这样做的好处是，由于 URL 中记录了页签的打开情况，通过 URL 便可恢复页签。

可使用如 URL 的 pathname 一级路径作为页签的标志，如果 URL 的一级路径为 foo，如地址为/foo/baz、/foo/biz，则激活 ID 为 foo 的页签。同理，如果地址为/boo/foz、/boo/biz，则激活 ID 为 boo 的页签。这里的标志不固定，可使用整个 pathname 部分，或者自行定义等。

本节将实现页签高阶组件，如图 9-5 中的 TabsProvider，其读取页签固定配置 TabsConfig，根据页面地址匹配页签数据，并将页签数据注入页签 UI 组件中。

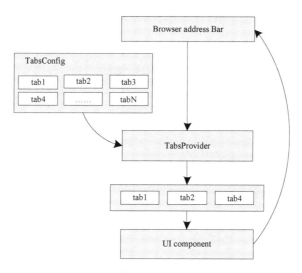

图 9-5　页签高阶组件运行示意图

9.10.2　页签配置

在页签系统中，页签也称为 tab，每个页签都有对应的配置，其配置类型如下：

```
export interface Tab {
  // tab 中记录的浏览器历史
  location?: Location;
  // 是否激活
  isActive?: boolean;
  // tab 的匹配模式，规定了页签所能匹配命中的 URL
  tabUrl: string;
  // 是否可关闭的 tab
  closable?: boolean;
  // 强制渲染的 tab
  renderWithOutMatched?: boolean;
  // tab 上的名字
  tabName: string;
  // 该 tab 的匹配函数，可选
  tabIsMatch?: (currentLocation: Location) => boolean;
}
```

location 的类型为第 2 章介绍的 history 中的 Location，拥有 pathname、search 等属性，可调用 history.createHref 生成对应的 URL 字符串。location 会在路由改变时更新，tab 会记录路由变化过程中最后一次匹配命中的路由。

isActive 标识该 tab 是否处于激活状态，每次路由变化时会计算该属性。

tabUrl 作为该 tab 的唯一标识，需要保证全局唯一，这里可设置为一级路径或者二级路径。若 tabUrl 为/foo，且 URL 以/foo 作为起始，则该 tab 应为激活状态。若没有额外配置 tabIsMatch 匹配函数，则所有以/foo 开头的路径都会激活此 tab。

closable 作为该 tab 是否可关闭的可选配置，可透传到 UI 组件中，UI 组件可根据配置执行相关逻辑。

renderWithOutMatched 作为可选配置，表示未命中时是否渲染 tab，用于某些"首页"性质的 tab。这类 tab 即便在路由未命中时也应进行渲染，其跳转路径为 location 指定的路径，且 location 也会根据路由变化情况进行更新。

tabName 表示页签的标题，类似于 document.title，一般表示该页签内页面的业务统称，如首页、商品和个人中心等。

tabIsMatch 是可选的匹配命中判断函数，如果不希望使用 tabUrl 作为起始 URL 的默认配置逻辑，则可传入此函数，如：

```
{
    tabName: "用户",
    closable: true,
    tabUrl: "/user",
    tabIsMatch: location => {
      return (
        location.pathname.startsWith("/user/list") ||
        location.pathname.startsWith("/user/detail")
      );
    }
}
```

以 user/list 或 /user/detail 开头的 URL 都会激活此 tab。

9.10.3 页签实现

页签系统的实现思路是，在高阶页签组件内维护有 2 个数组，一个数组为该页签组件能渲染出的所有的页签集合，通过 props.allTabs 传入，类型为 Tab[]；另一个数组初始为空，用于保存实际应该渲染出的页签。在每次 URL 改变时，会根据 currentLocation、tabUrl 或 tabIsMatch 计算页签的命中激活状态及是否应将页签加入页签渲染列表中。

与此同时，每次判断页签激活命中后，应该更新 location，这样在未来的页签切换时才可导航到页签内的最后一次操作页面。

此外，还需梳理页签关闭的两种场景。

1. 关闭当前激活项

当前激活项关闭，即关闭当前 tab。在浏览器中，如果关闭一个原生页签，则浏览器会取关闭原生页签右边位置的页签进行激活。如果是最后一个原生页签，则会取该原生页签左侧的页签进行激活。某些业务也会选择关闭激活 tab 后跳转到如"首页"的 tab，而无论被关闭的 tab 位于何处位置。本节使用了前者实现，首先通过 findNearestTab 找到距离被关闭 tab 最近的 tab，并将被关闭 tab 从渲染数组中移除，再通过 history.push 跳转到最近 tab 的 location 位置。

2. 关闭未激活 tab

关闭未激活的 tab，当前的 URL 应该不变，同时在渲染列表中应移除该 tab。注意，关闭未激活的 tab，应当移除该 tab 下页面的 DOM 元素，在实现上可通过调用 history.replace(currentLocation, state)的形式，不改变当前的 URL，并触发一次渲染。通过 state 传递消息的形式通知到对应的组件中，组件可根据 state 的属性值销毁 DOM 元素。

对于页签的激活，可通过 tabMatch 判断，有两种激活方式：

第一种，tab 定义时自行传入了 tabIsMatch 方法，将使用该方法将当前 location 传入，由业务侧自行判断 tab 是否匹配。

第二种，如果 tab 配置没有传入 tabIsMatch 匹配函数，则提供默认的匹配函数。默认的配置函数可以在创建 TabProvider 时自行传入，如果创建 TabProvider 时不传入默认的匹配函数，则使用 location.pathname 以 tabUrl 开头的默认匹配逻辑：

```
// 默认匹配判断函数，规定了以什么标准作为页签路由
function defaultTabMatch(tabUrl: string, location: Location) {
  return location.pathname.startsWith(tabUrl);
}
```

用于 tab 的激活判断：

```
// 匹配判断函数
function tabMatch(tab, currentLocation, defaultMatch = defaultTabMatch) {
  if (tab.tabIsMatch) {
    return tab.tabIsMatch(currentLocation);
  }
  return defaultMatch(tab.tabUrl, currentLocation);
}

// tabMatch 的使用
  const activedTabwithinRendered = renderdTabs.find(tab =>
    tabMatch(tab, currentLocation)
  );
```

TabsProvider 高阶组件的实现如下：

```
interface Props {
  allTabs: Tab[];
  history: History;
  children: React.ReactElement;
```

```
}
export default function(props: Props) {
  const { location: currentLocation } = React.useContext(__RouterContext);
  // 用于渲染 tabPane 的数组
  const renderdTabsRef = React.useRef<Tab[]>([]);
  // 复制一份用于渲染的 tabs
  const renderdTabs: Tab[] = [...renderdTabsRef.current];
  // 根据当前路径查找渲染 tab 中匹配成功的
  // 用于在已渲染的 tabpane 中判断激活项
  const renderedTab = renderdTabs.find(tab => tabMatch(tab, currentLocation));
  // 从所有的配置项中找寻与路径匹配成功的
  const shouldActivedTab = props.allTabs.find(tab =>
    tabMatch(tab, currentLocation)
  );
  // 清空激活态
  renderdTabs.forEach(renderTab => (renderTab.isActive = false));
  if (renderedTab && shouldActivedTab) {
    //激活项在已渲染的项内
    renderedTab.location = currentLocation;
    renderedTab.isActive = true;
  }
  // 渲染项存在，但是外部配置将该渲染项移除
  // 渲染项如[a,b,c]，但是外部配置从[a,b,c,d] -> [b,c,d,e]
  // 则渲染项也对应移除
  if (renderedTab && !shouldActivedTab) {
    const tabIndex = renderdTabs.indexOf(renderedTab);
    // 移除不在配置中的渲染项
    renderdTabs.splice(tabIndex, 1);
  }
  if (!renderedTab && shouldActivedTab) {
    renderdTabs.push({
      location: currentLocation,
      isActive: true,
      ...shouldActivedTab
    });
  }
  // renderWithOutMatched 配置通常用于"首页"的 tabpane，页签的渲染与 URL 的命
  // 中与否无关，强制渲染项不需要命中就能将 tabPane 渲染出来
```

```
const mainTabs = findAll(
  props.allTabs || [],
  tab => tab.renderWithOutMatched
);
if (mainTabs && mainTabs.length) {
  // 如果当前的渲染中没有强制渲染项，则将该项加入渲染
  renderdTabs.unshift(..._.differenceBy(mainTabs, renderdTabs, "tabUrl"));
}
// 更新渲染的值
renderdTabsRef.current = renderdTabs;
// 传入 closeTab 函数给子组件，当子组件关闭页签时调用此函数
function closeTab(tab: Tab) {
  if (renderdTabs.length <= 1) return;
  const tabIndex = renderdTabs.indexOf(tab);
  if (tabIndex !== -1) {
    if (!tab.isActive) {
      // 关闭的 tabpane 为非激活的 tabPane
      renderdTabs.splice(tabIndex, 1);
      // 使用 props.history 方法，具体行为由 props.history 控制
      props.history.replace(currentLocation);
    } else {
      // 关闭的 tabpane 为当前激活的，需要找到距离关闭项最近的 tabpane 进行激活
      const nearestTab = findNearestTab(tab, renderdTabs);
      renderdTabs.splice(tabIndex, 1);
      props.history.push(nearestTab.location);
    }
  }
}
// 传入 activeTab 函数给子组件，在子组件切换页签跳转时可调用此函数
function activeTab(tab: Tab) {
  // 使用 props.history 方法，具体行为由 props.history 控制
  props.history.push(tab.location);
}
return React.cloneElement(React.Children.only(props.children), {
  tabProp: { closeTab, activeTab, tabs: renderdTabs }
});
}
```

拥有 renderWithOutMatched 属性的 tab 也可理解为需要强制渲染的 tab，而与具体的 URL 路

径无关。在实现时，在组件内判断是否有强制渲染项，通过取差集的形式获取到应渲染的部分进行渲染。

如果希望能改写默认的 tab 匹配策略，则也可包装高阶函数，提供 createTabsProvider，并在创建 Provider 时提供默认的 tab 匹配策略：

```
const TabsProvider = createTabsProvider({
  defaultMatch: (tabUrl: string, location: Location) => {
    // 默认的 tab 匹配策略
    return location.pathname.startsWith(tabUrl);
  }
});
    // 渲染 Provider
    // 运行时传入 allTabs，好处是可以动态变更所有的渲染项
<TabsProvider allTabs={allTabs}>
    <Tabs>{props.children}</Tabs>
</TabsProvider>
```

通过父组件 TabsProvider 传入属性后，相关页签 UI 子组件可通过 props.tabProp 读取到 tabs 的信息：

```
// Tabs 组件示例
// tabs.tsx 负责 UI 渲染的 Tabs 组件
export default function(props) {
  if (!props.tabProp || !props.tabProp.tabs) return null;
  return (
    <div className="tabs">
      <div>tab 内容区域</div>
      <div className="tabs-nav">
        {props.tabProp.tabs.map((tab, index) => {
          return (
            <TabItemHeader
              key={index}
              active={tab.isActive}
              onJump={() => {
                // 单击 tab 时激活此 tab
                props.tabProp.activeTab(tab);
              }}
              onClose={() => {
                // 关闭 tab
```

```
            props.tabProp.closeTab(tab);
          }}
          closable={tab.closable}
          title={tab.tabName}
        />
      );
    })}
    </div>
      {props.children}
    </div>
  );
}
```

9.11 在 React Hooks 中使用路由

9.11.1 通过 React Hooks 获得路由组件

在 React v16.8 及以上版本中，提供了原生 Hooks。Hooks 在使用得当的情况下，可大大减少代码的行数，同时 React 也支持自定义 Hook。如若通过自定义 Hook 声明路由，如提供 useHookRoutes 自定义 Hook，则可以快速地实现小型应用，如以下示例：

```
import { useParams } from "react-router-dom";
import useHookRoutes from "./hookrouter";
import Detail from "./Detail";
// 声明路由路径及组件
const routes = {
  "/list": List,
  "/detail/:id": {
    component: Detail,
    routeConfig: {
      exact: true
    }
  }
};
export default function(props) {
  // 自定义 Hook，返回应该渲染的组件
```

```
  const component = useHookRoutes(routes);
  const params = useParams();
  React.useEffect(() => {
    /* ............ */
  }, []);
  return (
    <>
      {params.id}
      {component || <NotFind />}
    </>
  );
}
```

上例 component 为 useHookRoutes 的返回值，当命中"/list"路径时，component 为 List 组件实例化的元素；当命中"/detail/123"路径时，component 为 Detail 组件实例化的元素；当渲染各组件时，由路由注入的 props 与使用 Route 渲染时一致，都为 RouteComponentProps 类型。如果都未命中路由，则 component 为 null，这时可以返回 NotFind 组件，对 404 页面的支持非常直观。

由于使用了 Hooks 定义组件，因此也可以方便地使用 useHistory、useLocation、useParams、useRouteMatch 等 React Router 中提供的 Hooks，如：

```
export default function(props: RouteComponentProps) {
  // 自定义 Hook，返回应该渲染的组件
  const component = useHookRoutes(routes);
  // 方便使用多种 Hooks
  const history = useHistory();
  // 获取当前地址
  const location = useLocation();
  // 获取命名参数
  const params= useParams();
  // 某些自定义 Hook
  const update = useUpdateQuery();
  return (
    <>
      {component || <NotFind/> }
    </>
  );
}
```

9.11.2 实现 useHookRoutes

useHookRoutes 为自定义 Hook，可在其中使用 React.useContext 获取 React Router 的 RouterContext，进而可以从 RouterContext 中获取到组件树中最近一级 Route 的匹配情况，以及当前的地址。获得当前地址后，可使用 React Router 提供的 matchPatch 方法，手动计算一次路径配置对象 routes 定义的各个路径的匹配情况。在计算路径匹配时，路径由上一级 Route 的命中路径与当前对象定义的路径拼接构成，这样可形成嵌套路径定义。

```
const routerContext = React.useContext<RouteComponentProps>(__RouterContext);
if (!routerContext) return null;
// 获取上下文中的命中情况
const { match: parentRouteMatch, location } = routerContext;
// 是否有一个路径能命中当前地址
const isSomeRouteMatched = Object.keys(routes).some(
  path =>
    !!matchPath(location.pathname, {
      path:
`${processPath(parentRouteMatch.path)}${processPath(path)}`,
      exact: false
    })
);
```

手动计算一次路径匹配情况，目的是判断是否所有定义的路径都未匹配成功，如果通过 isSomeRouteMatched 判断没有一个路由路径匹配成功，则 useHookRoutes 可返回 null，以便使用该 Hook 的组件渲染未匹配页面，如 NotFind 页面。

在路径配置 routes 中，也可以映射路径为对象，并可从中获取到 routeConfig，其类型与 Route 的 props 配置类型 RouteProps 一致，这样可对 Route 提供更精细的配置，如配置 exact、strict 属性等。完整 useHookRoutes 的实现如下：

```
interface HookRoutes {
  [key: string]:
    | React.ComponentType
    | { component: React.ComponentType; routeConfig: RouteProps };
}
export default function useHookRoutes(routes: HookRoutes) {
  // 获取路由上下文
  const routerContext = React.useContext<RouteComponentProps>(__RouterContext);
  if (!routerContext) return null;
```

```js
const { match: parentRouteMatch, location } = routerContext;
// 手动计算判断，用于返回 null 的情景
const isSomeRouteMatched = Object.keys(routes).some(
  path =>
    !!matchPath(location.pathname, {
      path: `${processPath(parentRouteMatch.path)}${processPath(path)}`,
      exact: false
    })
);
// 匹配得到了 ReactElement，用于返回
const element = parentRouteMatch && (
  <Switch>
    {Object.entries(routes).map(([path, component], index) => {
      let routeConfig;
      let Component = component;
      if (typeof component === "object") {
        routeConfig = component.routeConfig;
        Component = component.component;
      }
      // 获取上下文中的路径，构成嵌套路径
      const pathRoute = `${processPath(parentRouteMatch.path)}${processPath(
        path
      )}`;
      return (
        <Route
          exact={false}
          key={index}
          path={pathRoute}
          render={(routeProps: RouteComponentProps) => {
            return React.createElement(Component, routeProps);
          }}
          {...routeConfig}
        />
      );
    })}
  </Switch>
);
// 在未匹配路由的情况下，返回 null，提供业务侧判断
```

```
if (!isSomeRouteMatched) {
  return null;
}
return element;
}
```

上例在计算 element 时，需要保证 parentRouteMatch 父级路由命中。在一般情况下，组件树中的某一级 Route 如果未命中，则该 Route 定义的页面不应展示，嵌套路由子路由应该跟随父路由的路径命中情况，所以该 Route 的子 Route 也同样不应渲染。因此，这里根据上级 Route 的命中情况，选择性地渲染了本级 Route。

渲染部分使用了 Switch，保证仅有一个路由路径得到渲染，将路由对象定义的路径与组件拆解到 Route 路由中，使用 Route 作为 Switch 的子组件渲染。同样，路径由上一级 Route 的命中路径与对象定义的路径拼接构成，拼接而成的路径将使得路由配置成为嵌套路由配置。在渲染方式上，Route 使用 render 渲染的方式，这样仅在 Route 匹配命中时才调用 render 函数渲染对应的组件，未匹配命中的 Route 返回 null。

最后，可根据 isSomeRouteMatched 是否有命中结果来返回 element 元素或返回空。

由于在本质上使用了 Route，因此 Route 所具备的一切能力 useHookRoutes 都能提供，如从组件的 props.match 中获取路径命中情况，或者从 props.match.params 中获取命名参数，以及从 props.location 中获取当前路径等。同时得益于 Hooks，使用自定义 useHookRoutes 的组件都可以自由使用 React Router v5.1.2 及后续版本中的各个自定义 Hook，如 useLocation 和 useHistory。

9.12　微服务路由

9.12.1　微服务介绍

近年来，前端单体架构正在过渡到许多较小、较易管理，可按服务进行拆分的前端架构，也称为微前端架构。微前端架构的好处主要是可以独立开发、独立部署、独立运行且与技术栈无关，比较适合新旧项目整合或业务繁多复杂的情况。

在一般情况下，微前端架构的实现有一个主项目作为调度中心，其功能是管理子项目，包括控制子项目加载、渲染、销毁等。本节将主要介绍主项目中的路由部分，包括获取子项目所注册的路由配置、按照配置加载子项目、响应外部地址等。主项目按路由加载子项目如图 9-6 所示。

第 9 章 进阶实战案例

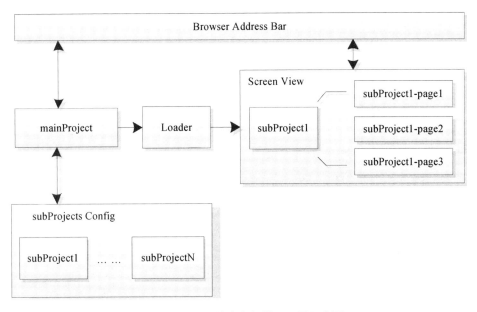

图 9-6 主项目按路由加载子项目示意图

9.12.2 实现示例

为了说明微服务的运作模式，本节引入一个示例场景。

1．背景

假设现有 3 个旧项目：
- 一个旧有单页 Vue 项目。
- 一个旧有单页 Angular 项目。
- 一个旧有单页 React 项目。

现在需要新开发组成一个大的单页 SPA 项目，除了包括以上 3 个项目，同时还需要新增功能模块。

2．目标

各个子项目可以独立开发、独立上线，各个子项目互不影响，子项目的路由形式也不关心，hash 路由或 browser 路由都可以，且子项目如有改造，改造量应尽可能小。

287

3. 实现思路

为了能很好地复用旧有项目，可考虑引入一个主项目来管理各个旧有项目和新增模块。为了说明 React Router 在微服务中的应用，这里的技术选型方案暂不涉及，不考虑 iframe 内嵌，以及 Angular、Vue 等方案。在技术选型上考虑使用 React。

作为管理性质的主项目，在业务上没有任何功能，其主要功能是根据各自业务配置拉取各自子项目、协调各子项目运作等。

加上原有的 3 个旧有项目及新增 React 项目与主项目，目前得到 5 个项目模块：

- 一个旧有单页 Vue 子项目。
- 一个旧有单页 Angular 子项目。
- 一个旧有单页 React 子项目。
- 一个新的 React 子项目，用来开发新需求新模块。
- 一个新的 React 主项目，用于管理各个子项目。

4. 设计实现

1）子项目注册

这一步的目的主要是让主项目提前获知子项目的信息，以便管理其加载、展示和隐藏等。子项目注册的过程不在子项目运行时进行，一般在开发阶段进行确认，即需要子项目提前通过某种方式告知主项目其子项目信息，如子项目将项目信息保存在数据库，主项目通过接口拉取等。子项目的配置可为如下类型的可序列化对象：

```
interface subProjectConfig extends RegisterConfig {
  // 项目全局唯一
  projectName: string;
  resources: any;
  shouldUnmountWhenLeave?: boolean;
  routerType?: "browser" | "hash";
  urlIdentifier?: string;
  containerId?: string;
}
```

projectName 为子项目的名字，这个名字全局唯一，如果遇到相同的项目名，则也应报错提示开发者。

resources 为子项目的资源地址，如 CSS、JavaScript、HTML 等，通常由主项目拉取这类资源。主项目拉取子项目资源本节不进行重点介绍，我们会重点关注路由部分。

shouldUnmountWhenLeave 为子项目的移除策略，如果为 true，则子项目在路由未命中时会被销毁。

routerType 表明子项目的路由类型，一般有浏览器路由与 hash 路由。

urlIdentifier 表明 URL 中能确定子项目的标志，如果为空，则可使用 projectName 作为 urlIdentifier，这种模式决定了子项目的激活即为通过路由激活，路由模式也相应地为被动式路由，即在运行时才能确认当前 URL 所应激活的子项目。

containerId 为子项目的容器 ID，用于给子项目提供挂载点。

2）运行机制

这里暂不考虑各个项目的构建、组件复用等，仅介绍各项目运行时的功能。

各子项目需提供能唯一标识自身的路由，如通过 basename 或其他方法。如果某子项目路由的 basename 为/foo 且不与其余项目冲突重名，则可认为/foo 唯一标识了该子项目。

各子项目可独立使用路由系统。无论是 hashHistory、browserHistory，还是 Vue、Angular、React，其路由运作的底层机制都一致，如浏览器基础方法 pushState、popstate 事件、hashchange 事件等。

主项目的职责是管控各个子项目的生命周期，如初始化、挂载、销毁，且保证屏幕中永远仅有一个子项目得到展示，并提供给各子项目跨项目跳转的机制，即跨项目跳转时，原项目在屏幕中消失，跳转的项目激活显示在屏幕中。

子项目为动态路由，主项目不需要知道子项目的路由数量，只需知道子项目的路由标识符。

3）各部分实现

主项目在初始化时执行一次 matchSubProject(history.location,subProjects)，目的是判断页面初始化时路由的匹配情况。其中，matchSubProject 函数的作用是判断各个子项目与传入参数的 location 是否匹配：

```
function matchSubProject(
  currentLocation: Location,
  // 维护了各个子项目的配置
  projectConfigs: subProjectConfigUsedByMainProject[]
```

```js
) {
  projectConfigs.forEach(config => {
    const uriIdentifier = config.urlIdentifier || config.projectName;
    // 判断 URL 是否匹配
    if (
      config.routerType === "hash" &&
      currentLocation.hash.includes(uriIdentifier)
    ) {
      // 设置匹配标志
      config._isMatch = true;
      return;
    }
    if (currentLocation.pathname.includes(uriIdentifier)) {
      config._isMatch = true;
      return;
    }
    config._isMatch = false;
  });
}
```

判断依据是 urlIdentifier，如果没有 urlIdentifier，则使用 projectName 进行判断，并根据 routerType 子项目的路由类型缩小判断的范围；如果是 browser 路由，则判断 window.location.pathname 部分；如果是 hash 路由，则判断 window.location.hash 部分。对每个匹配成功的子项目设置 _isMatch 标志，以便后续使用。

主项目在初始化后使用 browserHistory 监听 history 的变化（能同时监听 browser 路径与 hash 路径），并在回调函数中再次调用 matchSubProject(history.location, subProjects)，以便地址变化时更新各个子项目的命中状态：

```js
const history = createBrowserHistory({ basename: "/base/foo" });
history.listen(location => {
  matchSubProject(location, subProjects);
});
function App() {
  return <Router history={history}>{renderSubProject(subProjects)}</Router>;
}
```

除了监听路由的变化，主项目作为 React 应用，还将渲染 React 组件，并需要渲染各个子项目的容器 DOM 元素，且需要在主项目应用的 update 阶段，根据各子项目的路由命中情况及各子项目的历史加载情况确定是否需要加载子项目。关于主项目拉取子项目，一般有两种模式。一种是拉取子项目的 HTML，设置到某元素的子节点下，并设置 mounted 标签，避免反复调用 bootstrap 拉取。另一种是加载子项目的 JavaScript 文件，主项目提供容器供子项目挂载。

两种模式主项目都应支持，可提供<div id={id} />的容器。这里提供了一个加载器组件示例，省略了 bootstrapSubProjectIfNeeded 的实现：

```
function SubProjectLoader(props: subProjectConfigUsedByMainProject) {
  const id = props.containerId || props.projectName;
  React.useEffect(() => {
    // 内部不会每次都执行 bootstrap，如果需要，则拉取子项目
    bootstrapSubProjectIfNeeded(props);
    mountSubProjectIfNeeded(props);
    unmountSubProjectIfNeeded(props);
  });
  // 主项目加入容器的 CSS 控制
  return <div id={id} style={{ display: props.isMatch ? "block" : "none" }} />;
}
```

在拉取成功后，再根据子项目的运行模式决定是否需要挂载，如果子项目为主动挂载的，则这里可不需要调用其挂载方法；如果子项目通过如下示例注册函数：

```
interface RegisterConfig {
  mount?: () => void;
  unmount?: () => void;
}
function registerSubProject(
  projectName: string,
  registerConfig: RegisterConfig
) {
  // 注册函数示例
  window.sandbox[projectName] = registerConfig;
}
// 子项目在其入口调用 registerSubProject，提供给主项目挂载和销毁能力
// 如某子项目在其入口调用：
// registerSubProject("user-center",{mount:()=>{/*……*/}})
```

在运行时注册了挂载方法，且没有自行主动挂载，主项目就需要调用子项目的挂载方法，如 mount 方法，进行子项目的挂载。另外，子项目在页面切换后被销毁，又重新渲染时，也需

要调用挂载方法进行挂载。

主项目调用挂载方法挂载子项目这里也省略了具体的实现。下面提供了主项目挂载子项目的示例，仅作参考：

```
const mountSubProjectIfNeeded = (config: subProjectConfigUsedByMainProject) => {
  /*
    没有子项目 DOM 节点
    shouldRenderSubProject 一般用于记录对于已经渲染过的子项目是否需要再次渲染
    _isMatch 为子项目计算的匹配标志
  */
  if (shouldRenderSubProject(config) && config._isMatch) {
    // 调用子项目提供的方法
    window.sandbox[config.projectName].render &&
      window.sandbox[config.projectName].render();
  }
};
```

在 effect 中，若子项目不需要挂载，则还需判断子项目是否需要被销毁。如果需要被销毁，则应调用子项目注册的 unmount 等函数，示例如下：

```
const unmountSubProjectIfNeeded = (
  config: subProjectConfigUsedByMainProject
) => {
  if (
    // shouldunMountSubProject 一般用于记录对于销毁过的子项目，如果未命中，
    // 是否再次需要调用
    shouldunMountSubProject(config) &&
    !config._isMatch &&
    config.shouldUnmountWhenLeave
  ) {
    window.sandbox[config.projectName].unmount &&
      window.sandbox[config.projectName].unmount();
  }
};
```

由于各个子项目拥有唯一的 URL 标识，因此当出现多个子项目命中一个 URL 的情况时，可检测并提示：

```
const matchedProjects = projectConfigs.filter(_ => _._isMatch);
invariant(
```

```
matchedProjects.length <= 1,
`注意以下项目 URL 匹配冲突,请检查:
${matchedProjects.map(
  config =>
    "|" + (config.projectName || "") + "-" + (config.urlIdentifier || "")
)}
`
);
```

对于各个子项目间的跨项目跳转,主项目提供给子项目跨项目跳转的接口,如各子项目可通过调用 window.sandbox.main.history.push("other_url")实现跨项目的跳转。为了实现跨项目间路由响应,主项目可对 browserHistory 进行包装提升,在 push、replace 方法中加入事件广播:

```
function enhanceHistory(history: History, methods = ["push", "replace"]) {
  const originHistoryMethods = {};
  methods.forEach(method => {
    originHistoryMethods[method] = history[method];
  });
  methods.forEach(method => {
    history[method] = (...args: any[]) => {
      originHistoryMethods[method](...args);
      if (typeof window.Event === "function") {
        // 为了触发一次路由变化广播,使各子项目响应。命中路由的子项目将接管页面控制权
        window.dispatchEvent(new HashChangeEvent("hashchange"));
        window.dispatchEvent(new PopStateEvent("popstate"));
      } else {
        // IE 浏览器情况
        const hashChangeEvent = document.createEvent("Event");
        hashChangeEvent.initEvent("hashchange", false, false);
        window.dispatchEvent(hashChangeEvent);
        const popstateEvent = document.createEvent("Event");
        popstateEvent.initEvent("popstate", false, false);
        window.dispatchEvent(popstateEvent);
      }
    };
  });
}
```

```
    return history;
}
```

当子项目调用如 window.sandbox.main.history.push 时，会触发各子项目路由系统内的底层订阅函数，如 popstate 或 hashchange 回调，由各子项目自动做出对应的响应。

由于子项目的路由类型未知，可 popstate 与 hashchange 事件都人工触发一次，分别对应 browser 路由与 hash 路由。

总结：从运行时的改造来说，子项目仅需保证自身有独立的路由标志，这可通过 basename 的方式实现，改造量较小、可控。

9.13 配置化路由扩展

9.13.1 配置化路由与 react-router-config

在 React Router v3.x 中，使用的是中心化静态路由，本质上路由都以配置形式存在，如以下示例：

```
const routes = [
  {
    component: Root,
    routes: [
      {
        path: "/user",
        component: A
      },
      {
        path: "/user/info",
        component: UserList,
        routes: [
          {
            path: "/user/info/:id",
            component: UserDetail
          }
        ]
      }
```

```
      ]
    }
];
export default routes;
```

路由配置以树状结构呈现，每一个节点都有 path、component 或可选的 routes 子路由属性。其中，path 为路由的路径，component 为所需渲染的组件，routes 为对应路由的子路由部分。使用了配置化结构的路由，由于每一级路由配置都有组件与之对应，路由树完全与组件树相对应，如图 9-7 所示。

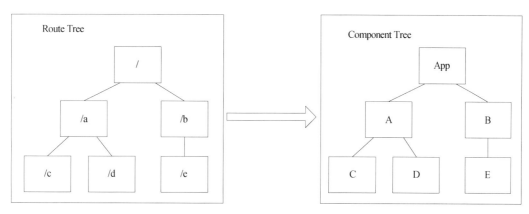

图 9-7　路由树对应到组件树

在 React Router v4.x 之后，由于路由变为了动态路由，React Router 不再支持静态的配置写法，若希望使用静态配置路由，一种方案是自行实现静态路由配置的渲染，另一种方案是使用 react-router-config 库提供的 renderRoutes 方法，用于简单处理配置化路由。如对于如上路由配置，可以使用 react-router-config 库中的 renderRoutes 方法进行组件渲染：

```
import { renderRoutes } from "react-router-config";
import routes from "./routes";
const Root = ({ route }) => (
  <div>
    Layout
    {renderRoutes(route.routes)}
  </div>
);
// 根路径
const A = props => <h2>标题</h2>;
```

```
const UserList = ({ route }) => (
  <div>
    {/* 渲染 route 子路由 */}
    {renderRoutes(route.routes)}
  </div>
);

const UserDetail = props => <span>user:{props.match.params.id}</span>;
const app = (
  <Router>
    {/* 引入定义好的路由配置，进行渲染 */}
    {renderRoutes(routes)}
  </Router>
);
```

renderRoutes 源码非常简单，如下所示：

```
import React from "react";
import { Switch, Route } from "react-router";
function renderRoutes(routes, extraProps = {}, switchProps = {}) {
  return routes ? (
    <Switch {...switchProps}>
      {routes.map((route, i) => (
        <Route
          key={route.key || i}
          path={route.path}
          exact={route.exact}
          strict={route.strict}
          render={props =>
            route.render ? (
              route.render({ ...props, ...extraProps, route: route })
            ) : (
              <route.component {...props} {...extraProps} route={route} />
            )
          }
        />
      ))}
    </Switch>
  ) : null;
}
export default renderRoutes;
```

源码中为每一个路由配置项的 routes 属性生成一个 Switch 组件及 Switch 组件管理的所有 Route 子组件。源码使用了 Route 的 Render Props 模式，使用了 render 属性进行渲染，渲染传入的 route.component 组件，或者调用 route.render 进行渲染。

从源码中还可以看到，route 还可以使用 key、exact、strict 和 render 属性。使用 key 的目的是保证 Route 组件在渲染过程中不进行不必要的销毁。因为 Route 组件都是作为 Switch 组件的子组件，Switch 组件保证 Route 子组件最终只渲染一个，若出现多个 Route 组件拥有相同的 key，却有着不同的路径，则在这些 Route 组件间进行路由切换时，对应的 Route 组件不会被销毁，只会执行组件更新的生命周期方法。

同时，react-router-config 还提供了 matchRoutes 帮助方法，用以获得路由配置项数组 routes 在某路径下的所有匹配结果。其特点是只匹配 routes 数组中第一个命中路径的 route，如使用本节开头所述的配置项，匹配路径/user/info：

```
import { matchRoutes } from "react-router-config";
const branch = matchRoutes(routes, "/user/info");
```

得到的结果如下：

```
[
  {
    route: routes[0],
    match: {
      path: "/",
      url: "/",
      params: Object,
      isExact: false
    }
  },
  {
    route: routes[0].routes[0],
    match: { path: "/user", url: "/user", isExact: false, params: Object }
  }
];
```

匹配结果中第一项的 route 为 routes[0]，即根路由/，根路由默认匹配命中。除根路由外，在上例中/user 匹配命中/user/info，因而匹配命中后跳过 routes 数组的后续路由配置项，得到上述结果。

9.13.2 重新实现配置化路由

上述 react-router-config 配置化路由存在一些问题，主要有：
- 切换页面 DOM 会被销毁，即无法使 DOM 得以保存，缺少缓存机制。
- 无法使用父子路由，path 需要书写完全路径。
- 无法使用 props.children 渲染子路由，渲染子路由需要人工引入 renderRoutes 并调用 renderRoutes(route.routes) 方法。
- 不支持 redirect 重定向。

综上所述，可对原配置进行扩展，提供如下新增配置属性：

```
{
  redirect?: string;
  standAlone?: string;
isIndexComponent?: boolean;
}
```

redirect 表明在匹配成功时需要重定向到的 URL 路径。

standAlone 表明组件的存在形式，如果为 true，则表明组件在未命中路由时不销毁，而是进行 CSS 隐藏处理，组件可被缓存。

isIndexComponent 表示是否为父配置提供默认的子组件，与 6.7.4 节所述的默认子组件 Route 类似。完整的 Route 配置项类型如下：

```
import { RouteProps } from "react-router";
export interface RouteExt extends RouteProps {
  redirect?: string;
  standAlone?: string;
  routes?: RouteExt[];
  isIndexComponent?: boolean;
}
```

经过改造实现的静态路由配置渲染方法，路由配置省去完全路径的写法，可以以父子相对路径进行声明：

```
const routeConfig = {
path: "/user",
component: UserList,
routes: [
  {
```

```
      path: "/info ",
      component: UserDetail
    }
  ]
};
```

对于子路由/info，组件为 UserDetail，父路由为/user，UserDetail 子组件对应到/user/info 才会渲染。

若父路由组件想要渲染子路由组件，则在需要渲染的位置返回 props.children 即可。

```
<div>
{props.children}
</div>
```

为了实现以上需求，可不使用 react-router-config 中的 renderRoutes 方法，重新实现 renderRoutes，调用方在顶层进行调用，如：

```
function App(routes: RouteExt[]){
  // routes 为路由配置数组
  return renderRoutes(routes)
}
```

设计 renderRoutes 如下：

```
import { RouteProps, Route, RouteChildrenProps, Redirect } from "react-router";
import { get } from "lodash";
import * as React from "react";
function renderRoutes(routes?: RouteExt[], parentPath?: string) {
  // 依次渲染子路由
  return routes && routes.map(route => renderRoute(route, parentPath));
}
```

调用 renderRoutes 函数的目的是渲染 routes 路由配置数组，函数对 routes 路由配置数组中的每一项路由配置调用 renderRoute 函数，主要的渲染工作在 renderRoute 中进行。为了弥补 Route 组件不能缓存页面的特性，renderRoute 函数使用了 6.7.5 节中介绍的 CacheRoute 实现缓存页面的能力：

```
export function addLeadingSlash(path) {
  return path.charAt(0) === "/" ? path : "/" + path;
}
export function stripTrailingSlash(path: string) {
```

```
    return path.charAt(path.length - 1) === "/" ? path.slice(0, -1) : path;
}
// 路径字符串处理
export function processPath(routePath?: string) {
    if (routePath === "/") return "";
    if (!routePath) return "";
    return addLeadingSlash(stripTrailingSlash(routePath));
}
const RenderChildren = props => props.children;
// 渲染配置
function renderRoute(route: RouteExt, parentPath?: string, renderUtil?: any) {
    const RouteComponent = route.component || RenderChildren;
    const currentPath = `${processPath(parentPath)}${processPath(route.path)}`;
    const currentRedirectPath = `${processPath(parentPath)}${processPath(
        route.redirect
    )}`;
    return (
        <Route
            // 透传属性
            {...omit(route, "component", "render")}
            // 如果 route 未配置 path，将根据父 Route 的命中情况进行渲染，在 path 为 undefined 的
            // 情况下，父路由配置命中则渲染，否则不渲染
            path={route.path && currentPath}
            // 如果是父路由，则 exact 需要为 false，防止出现子路由不渲染的情形
            exact={route.routes && route.routes.length ? false : route.exact}
            children={(routeProps: RouteChildrenProps) => {
/* CacheRoute 可查看 6.7.5 节 */
                return (
                    <CacheRoute
                        shouldDestroyDomWhenNotMatch={!route.standAlone}
                        shouldMatchExact={route.isIndexComponent}
                        render={routeProps => (
                            <>
                                {route.redirect &&
                                    routeProps.match &&
                                    routeProps.match.isExact && (
                                        <Redirect key={Math.random()} to={currentRedirectPath} />
                                    )}
```

```
            <RouteComponent
              {...routeProps}
              routePath={currentPath}
              route={route}
            >
              {!route.isIndexComponent && renderRoutes(
                route.routes,
                currentPath,
              )}
            </RouteComponent>
          </>
        )}
      />
    );
  }}
  />
  );
}
```

上例使用了 Route 与 CacheRoute，其中 Route 用于路径匹配，CacheRoute 继承 Route 的路由匹配结果，提供缓存支持。

在 Route 的 path 选择上，使用通过 renderRoutes 传递的 path 作为父 path，这样子路由便可拼接父路由 path 与自身的路由 path：

```
const path = `${parentPath}/${route.path}`
```

如父路由配置为/a，那么所有子路由的最终路径都将以/a 开头。

对于 Route 的 exact 属性，也需要设置。在配置拥有 routes 属性且长度不为 0（表示拥有子路由配置）时，exact 将为 false，Route 的匹配方式为模糊匹配。原因在于父子配置的特性，需要保证父配置 Route 的 exact 属性为 false。如果父配置 exact 为 true，则仅能精确匹配本级路径，不能模糊匹配后续路径。如在路径为/a/b 的路由配置下，Route 的 exact 为 true，则在外部地址为/a/b/c 时不会渲染/a/b 对应的组件。由于/a/b/c 作为/a/b 的子配置依赖父级/a/b 的渲染，而在/a/b得不到渲染的情况下，/a/b/c 也得不到渲染，与预期不符，子配置无法生效。所以，在上述实现中对 exact 做了判断，在拥有 routes 子配置时，exact 为 false，这保证了在渲染后续的子路由组件时，父路由组件也可得以渲染。由于路由配置呈现树状结构，因此可以理解为 exact 为 false的配置类似于树的非叶子节点，exact 为 true 的配置在路由树中为叶子节点。

在实现的 renderRoute 方法中，使用 Route 组件用于路径的匹配，并使用 children 属性无条件渲染 CacheRoute 子组件。由于 CacheRoute 作为特殊的 Route，继承拥有 Route 的特性，父级 Route 组件的路径匹配结果也将传入 CacheRoute，因而不必为 CacheRoute 传入 path 属性，父组件 Route 的路径匹配结果即为 CacheRoute 的路径匹配结果。此处，CacheRoute 使用了 Render Props 渲染模式，传入了 render 方法，渲染页面组件。

在业务组件中，可使用 props.children 渲染子路由组件，原因在于在路由配置项中 route.component 的 children 部分使用了 renderRoutes，props.children 即为 renderRoutes 的返回值。

对于路由配置，允许传入 redirect 字符串，redirect 的作用是对当前路由配置进行重定向操作。如果传入了 redirect 字符串，并且当前的路由匹配结果为绝对匹配，则可渲染 Redirect 组件，进行跳转。Redirect 组件的 key 在每次渲染时都不同，这保证了在满足重定向条件下 Redirect 组件都可重新挂载，进而执行重定向操作。Redirect 组件的渲染与 route.component 组件的渲染互相独立，互不影响，处于平级关系，这样保证了在 route.component 组件正常渲染且不被销毁的同时，也能渲染 Redirect 组件进行重定向操作。

配置 standAlone 属性，目的是使得路由配置所对应的页面能得到缓存，在页面的切换过程中，DOM 不被销毁。实现此方式本质上利用了 Route 中 children 的无条件渲染模式，并利用 props.match 判断命中情况，页面的隐藏通过 display:none 实现，此处复用了 6.7.5 节中 CacheRoute 的能力，将!route.standAlone 传入 CacheRoute 的 shouldDestroyDomWhenNotMatch 属性，相关介绍可查看 6.7.5 节。

在路由配置中，可设置 isIndexComponent。对于设置 isIndexComponent 为 true 的路由配置，可称为默认子组件路由配置，即在父配置路径绝对命中的情况下为父配置的组件提供默认的子组件，与 6.7.4 节的默认子组件 Route 能力相同。在 route.isIndexComponent 为 true 时，路由配置不需要注明 path 属性，其作用是为父级组件提供子组件。在代码实现上，由于默认子组件路由配置的 route.path 为 undefined，渲染的 Route 为无路径 Route（pathless Route），无路径 Route 会继承使用上级 Route 的命中结果。需要注意，默认子组件的配置的命中模式为全匹配命中，即需要判断 match.isExact 为 true，否则不满足父路径绝对命中的前提要求。在实现上，将 route.isIndexComponent 传入 CacheRoute 的 shouldMatchExact 属性，CacheRoute 内部将判断 match.isExact 是否为 true。若不将 route.isIndexComponent 传入 CacheRoute 的 shouldMatchExact 属性，默认子组件路由配置将会不满足父路径绝对命中的前提，将会有着与父配置一样的匹配模式，与父配置将无差异可言。

需要注意，默认子组件路由配置的 routes 属性也应忽略，由于默认子组件路由配置的匹配模式是精确匹配，默认为路由树的叶子节点，routes 的子路由不应存在。若要提供子配置，则需要舍去匹配的精确性，节点也将变成非叶子节点，但这与配置的叶子节点性质产生冲突。在实现上，route.isIndexComponent 为 true 的配置，不再调用 renderRoutes 渲染 children。

9.14　配置化路由综合示例

在中大型项目中，使用动态声明式路由存在一定的问题，如 Route 声明不规范、Route 嵌套层数过深、子 Route 全路径声明过于烦琐等。这类动态 Route 的使用虽然让项目开发变得灵活，但在缺失开发规范或者框架限制的情况下，大量使用动态路由会使得项目后期难以管理和维护，也会使得一些组件和页面拥有很多不必要的"副作用"。

因此，本节不使用组件化 Route 灵活的声明方式，将开发一个综合的配置化路由框架。路由框架将业务开发中的路由管理解放出来，业务侧无须关心 Route 的声明、父子路由、Redirect、缓存等问题，仅需按照简单配置方式实现路由定义。考虑 9.8 节 history 的导航中间件、9.10 节较为复杂的页签场景与 9.13 节中的配置化路由扩展，本节计划结合导航中间件、页签机制、路由配置等内容，为读者呈现一个综合性的带页签的页面示例。示例使用了常用的两栏式布局，布局中包含导航菜单，以及与导航菜单一一对应的页签，如图 9-8 所示。

图 9-8　页面布局示意图

在图 9-8 中，左侧导航菜单提供固定的导航支持，每个导航菜单对应一个页签，页签可如浏览器页签一样进行关闭。在一般情况下，应用的页签处于关闭状态，当单击导航菜单后，会激活菜单对应的页签，并打开该菜单对应的页面。通常一个页签下可以包含多个页面，一般为一个业务的完整流程所需的页面集合。同时，为了提供用户刷新页面机制，获得良好的交互体验，无论页签打开与否，以及页面状态如何，单击导航菜单都会跳转到菜单 URL 所对应的页面，并使页面区域内的页面内容无条件刷新。

9.14.1 路由配置

考虑到在浏览器 tab 中，用户的操作记录应停留在原 tab 中，为了模拟相同的功能，设计的路由框架需能缓存页面。

与此同时，在一些场景中，如单击左侧导航侧菜单，无论是单击同一个菜单还是切换其他菜单，希望的都是强制更新该菜单对应页面的状态，参考浏览器的刷新功能，其会丢失当前的用户操作记录，并重新发送请求加载最新的页面数据。通过提供页面内的强制刷新能力，能提供给用户较好的浏览体验而不需真正刷新浏览器。

同样，由于一个页签下仅有一个页面展示，因此页签内的页面作为一个小型 SPA 应用形态。在关闭页签时，需要考虑销毁页签内的 DOM 元素，因为在各个页签内的页面通常是隔离的页面，多数情况下在关闭页签时并不需要为用户保留页签的残留状态。

通过上述描述，框架应提供三类能力：记忆渲染能力、remount 能力和销毁组件能力。

为了实现上述三类能力，可在 9.13 节配置化路由的基础上适当加以扩展，无须额外开发渲染框架。

1. 状态缓存

在浏览器 tab 中，用户的操作记录停留在原 tab 中，如用户输入的文本、用户滑动滚动条所停留的位置等，这些操作的结果在切换页签后不会消失。为了模拟浏览器相同的功能，除了可使用 9.10 节中的高阶页签组件提供支持，模拟的页签机制还需要具备路由页面的缓存能力。

在 9.13 节中曾介绍过配置化路由的 standAlone 缓存能力，通过隐藏页面节点，可以做到页面缓存。由于 tab 的切换操作是一个高频率的操作，在框架实现上可将 standAlone 作为一个全局化配置，为所有页面全局引入：

```
let useStandAlone = false;
export function initAppType(options) {
```

```
    useStandAlone = options.useStandAlone;
}

// App.tsx
// 在应用入口导入 initAppType
initAppType({
  useStandAlone: true
});
```

在路由配置判断是否需要缓存页面时，仅需扩展实现 9.13 节的 destroyDom 判断：

```
const destroyDom = !(route.standAlone || useStandAlone);
```

便可使用此全局配置配置全局的页面缓存。

在通过 useStandAlone 配置了全局的页面缓存后，使用 history 导航。当调用 history 各方法时，各页面都将被缓存。如当前页面路径为/user/list，当调用 history.push("/user/detail")导航到/user/detail 后，原页面/user/list 将被缓存。这满足了页签场景的需求，在各页签间切换时，各页面状态都不会被销毁。

2．强制刷新

为了给用户提供良好的交互体验，在单击导航菜单时，无论页面状态如何，都将刷新该菜单对应的页面区域内容。由于路由应用为 SPA 应用，各个页面在实质上都是 React 组件，为了实现组件的刷新，可使用 React 的 remount 机制。在 6.7.5 节曾有过介绍，组件重新挂载常见的做法是在渲染 React 组件时为 React 组件设置不同的 key。在本节中，需要确定的是组件的重新挂载以何种方式得到触发。

考虑到 history 的 state 可以保存导航前后的状态，如果使用 state 作为载体，保存触发 remount 的信号，则不仅不需要引入额外的通信成本，而且从 Route 组件注入业务组件中的 routeProps.location.state 中也可获取 state 中的相关信息。且由于浏览器历史栈会持久化记录导航 state 的值，当单击浏览器的"前进"和"后退"按钮时也能重现 state 中触发 remount 的信号。与此同时，在触发信号的选择上，由于路由配置每一级都拥有 path 与之对应，可在 state 中设置需要触发的路由路径，如设置 state._option.remountPath 为/user/info，则表明完全匹配/user/info 的路由组件需要重新挂载。

在 remount 机制的实现上，可从 Route 渲染函数中读取 routeProps.location.state 中的_option.remountPath 值，如果当前路由配置路径与 remount 路径完全匹配，则需要对本级路由的组件进行重新挂载。由于 6.7.5 节中的 CacheRoute 支持重新挂载操作，因此可将 remount 信号传入 CacheRoute。CacheRoute 内部使用 useRef 保存组件的 key，当外部接收到 remount 信号时，更改 key 的值并保存到 Ref，并对组件使用此更新后的 key，利用 key 的变化使得组件重新挂载。

基于上述思路，可对 9.13 节的配置化路由进行扩展，在 children 渲染函数中引入：

```
// 获取 remount 路径
const remountPath = get(routeProps.location.state, "_option.remountPath");
// 判断路径是否与当前路由路径相同
// 如果当前的 path 与 remountPath 对应的值相同，则说明当前的 Route 配置所对应的组
// 件需要重新挂载
const shouldRemountComponent =
  // 使用 currentPath 与 remountPath 比对
  remountPath && pathEqual(currentPath, remountPath);
```

上例当获取 state 中的 remountPath 后，与当前路由配置的全路径 currentPath 进行路径比对：

```
import { matchPath } from "react-router";
export function addLeadingSlash(path) {
  return path.charAt(0) === "/" ? path : "/" + path;
}
export function stripTrailingSlash(path: string) {
  return path.charAt(path.length - 1) === "/" ? path.slice(0, -1) : path;
}
export function processPath(routePath?: string) {
  if (routePath === "/") return "";
  if (!routePath) return "";
  return addLeadingSlash(stripTrailingSlash(routePath));
}
// 路径比对函数
function pathEqual(pathA: string, pathB: string): boolean {
  pathA = processPath(pathA);
  pathB = processPath(pathB);
  const matched = matchPath(pathB, { path: pathA });
  // 需要完全匹配，即 isExact 为 true
  return matched && matched.isExact;
}
```

当比对成功时，shouldRemountComponent 为 true，表明需要重新挂载组件，这时可将 shouldRemountComponent 传入 CacheRoute 的 shouldReMount 中。

在业务侧进行组件的重新挂载，无须额外引入触发方法，当 history 的 state 传入 _option. remountPath 属性，值为需要重新挂载的路径时，对应路径的路由组件便会执行重新挂载操作。调用方式如下：

```
// /a/b/c 路径对应的组件会被销毁并重新渲染
history.push("/a/b/c", { _option: { remountPath: "/a/b/c" } });
```

注意，如果 remountPath 为：

```
history.push("/a/b/c", { _option: { remountPath: "/a" } });
```

这样会使得/a 路由配置所对应的组件重新挂载，因而/a 路径下的子组件也会被销毁并重新挂载。

3. 销毁机制

在 9.10 节中，曾讨论过页签关闭的两种场景：一种是关闭当前激活页签，另一种是关闭已存在但未激活页签，但 9.10 节未涉及如何销毁缓存页面。本节在路由渲染层面提供路由组件的销毁机制，使得页签在关闭时，在页签内的路由页面得到正确销毁。

9.14.1 节曾介绍通过指明 history 的 state 中的 remount 路径使得对应路由组件得以重新挂载，同理，对于缓存页面的销毁，也可使用类似的方法：

```
// 读取 history.state 中的 unmountPath 值，判断是否需要销毁此路由组件
// 在路由未匹配的情况下
const unmountPath = get(routeProps.location.state, "_option.unmountPath");
const shouldUnmountComponent =
  unmountPath && pathEqual(currentPath, unmountPath);
```

在 6.7.5 节中曾介绍了 shouldDestroyDomWhenNotMatch 属性，若设置 shouldDestroyDomWhenNotMatch 为 true，则 CacheRoute 在路由未匹配命中的情况下，将不渲染组件，返回 null，与原生 Route 能力一致。对于 CacheRoute 曾经缓存过的页面组件，若 shouldDestroyDomWhenNotMatch 为 true，在未匹配路由时返回 null，意味着组件被销毁，这也是所需的销毁组件机制。在实现上仅需将销毁信号 shouldUnmountComponent 传入 CacheRoute，由 CacheRoute 负责组件生命周期的管理工作。

结合上一节的 remount 机制，最终改写后的路由配置渲染方法如下：

```
export interface RouteExt extends RouteProps {
  redirect?: string;
  standAlone?: string;
```

```
  routes?: RouteExt[];
  isIndexComponent?: boolean;
}
// 渲染 RouteExt 配置
function renderRoute(route: RouteExt, parentPath?: string, renderUtil?: any) {
  // 配置中的组件
  const RouteComponent = route.component || (props => props.children);
  // 拼接当前路径
  const currentPath = `${processPath(parentPath)}${processPath(route.path)}`;
  // 拼接重定向路径
  const currentRedirectPath = `${processPath(parentPath)}${processPath(
    route.redirect
  )}`;
  // 本级配置是否拥有默认子组件配置
  const hasChildrenIndexRoute =
    route.routes && route.routes.find(route => route.isIndexComponent);
  // 判断渲染模式是缓存还是销毁，standAlone 表示缓存路由
  const destroyDom = !(route.standAlone || useStandAlone);
  return (
    <Route
      // 透传属性
      {...omit(route, "component", "render")}
      // 如果 route 未配置 path，将根据父 Route 的命中情况进行渲染，在 path 为
      // undefined 的情况下，父路由配置命中则渲染，否则不渲染
      path={route.path && currentPath}
      // 如果是父路由，则 exact 需要为 false，防止出现子路由不渲染的情形
      exact={route.routes && route.routes.length ? false : route.exact}
      children={(routeProps: RouteChildrenProps) => {
        // 读取 history.state 中的 remountPath 值，判断是否需要重新挂载此路由对应组件
        const remountPath = get(
          routeProps.location.state,
          "_option.remountPath"
        );
        // 读取 history.state 中的 unmountPath 值，判断是否需要销毁组件
        // 在路由未匹配的情况下
        const unmountPath = get(
```

```
      routeProps.location.state,
      "_option.unmountPath"
    );
    // 如果当前的 currentPath 匹配 remountPath 对应的值，则说明当前的 Route
    // 配置所对应的组件需要重新挂载
    const shouldRemountComponent =
      // 使用 currentPath 可以兼顾 route.path 为 undefined 的情形
      remountPath && pathEqual(currentPath, remountPath);

    const shouldUnmountComponent =
      unmountPath && pathEqual(currentPath, unmountPath);

    return (
      <CacheRoute
        // 当 state 传入了 remount 信号，并且没有默认的子组件时
        //（拥有默认子组件，应该对默认子组件执行 remount 操作）才应该重新挂载
        shouldReMount={shouldRemountComponent && !hasChildrenIndexRoute}
        // 判断 standAlone 或者 state 中的 unmount 销毁信号，并且没有默认的
        // 子组件（拥有默认子组件，应该对默认子组件执行 unmount 操作），
        // 都满足条件时应该销毁
        shouldDestroyDomWhenNotMatch={
          (destroyDom || shouldUnmountComponent) && !hasChildrenIndexRoute
        }
        // 默认子组件应完全匹配
        shouldMatchExact={route.isIndexComponent}
        // Render Props 渲染模式
        render={routeProps => (
          <>
            {route.redirect &&
              routeProps.match &&
              // 在完全匹配的情况下重定向生效
              routeProps.match.isExact && (
                // 保证 Redirect 的渲染
                <Redirect key={Math.random()} to={currentRedirectPath} />
              )}
            {/* 渲染业务组件 */}
            <RouteComponent
```

```
              // 传入 routeProps
              {...routeProps}
              routePath={currentPath}
              route={route}
            >
              {/* 渲染子路由 */}
              {renderRoutes(
                route.routes,
                // 作为父路由传入
                currentPath,
                route.isIndexComponent
              )}
            </RouteComponent>
          </>
        )}
      />
    );
  }}
/>
);
}
```

9.14.2 导航

1. 四类导航

在本节的应用示例中，除了在页签间进行导航切换时需要保存页面信息，在关闭页签时，也需要模拟浏览器跳转到其他页签下。同时，在页签外层的导航菜单中进行导航应该强制刷新页面，且由于一个页签内可存在多个页面，也应该有页签内页面间的跳转导航。对于这些跳转，它们应有不同的表现形式，如页签间切换、页面应缓存、导航菜单的单击、对应页面应该刷新等。

通过上述描述及梳理 9.14.1 节中介绍的应用布局及交互方式，本节的应用示例应该拥有四类导航：菜单导航、页签间切换导航、关闭页签导航和页签内页面间导航。四类导航应有不同的调用方式，本节先设计四类导航接口，再结合上节中提供的 state._option.remountPath、state._option.unmountPath 能力，综合实现各类导航。

1）菜单导航

这类菜单导航类似于在浏览器中新开 tab，或者通过单击菜单切换到某个已存在的 tab 上。菜单导航独立于页签机制，其特点是通过菜单导航的页面都会强制刷新。单击菜单导航，特点是该菜单对应的页面应该重新渲染，菜单所对应的页面应执行一次重新挂载，同时其他菜单所对应的页面及页签所对应的页面不受影响，DOM 应该保留缓存。如上分析，为 history 的 push、replace 引入新的函数签名：

```
interface historyOptionExt {
  preserveDom?: boolean;
  forceRemount?: boolean;
}
push(path: string, state?: any, option?: historyOptionExt): void;
replace(path: string, state?: any, option?: historyOptionExt): void;
```

则菜单导航调用方式如下：

```
history.push(path, state, {
  // 强制跳转 path 重新渲染
  forceRemount: true,
  // 保留其他 DOM
  preserveDom: true,
});
  // 或者
<Link forceRemount preserveDom to= "/foo">
```

2）页签间切换导航

类似在浏览器中切换各个 tab，这类导航会将页面状态进行缓存。页签切换应该模拟浏览器页签行为，各页签内的内容 DOM 元素应该被保留。使用参数列表中最后的参数 option，接口设计如下：

```
history.push(path, state, {
  // 不强制重新挂载
  forceRemount: false,
  // 所有的 DOM 应该保留
  preserveDom: true,
});
  // 或者
<Link preserveDom to= "/foo">
```

3）关闭页签导航

这类导航将会在页签关闭的同时，导航跳转到其他页签下，如浏览器的页签在关闭后会跳转到被关闭页签的右侧页签。此类导航由于页签被关闭，对应页签下的页面应该被销毁。被关闭的页签内的页面内容应该被销毁，且关闭页签跳转到其他页签后，其他页签内的内容应不受影响。可使用 state._option.unmountPath 销毁页面：

```
// 销毁存在的非激活页签
history.replace(history.createHref(currentLocation), {
  _option: {
    // 销毁页签路径所对应的组件
    unmountPath: tab.location.pathname
  }
});
// 销毁当前激活页签
history.push(history.createHref(nearestTab.location), {
  _option: {
    // 销毁页签路径所对应的组件
    unmountPath: tab.location.pathname
  }
});
```

仅使用了 state._option.unmountPath 进行目标页面的销毁，对其余页面不造成影响。

4）页签内页面间导航

在通常情况下，页面区域仅能展示单个页面内容，而一个页签下一般拥有多个页面。当在这些页面间导航跳转时，除了导航目标页面需要挂载，可销毁不在页面区域中的页面。针对单个页签内有多个页面的情况，一个页签内同一时刻仅有一个页面得到渲染展示，页签内未命中路由的页面应该销毁，命中路由的页面应该渲染展示。此调用在业务开发中最为频繁，因此此调用不用传递其他参数，在设计上与原始 history 的调用方式一致：

```
history.push("/baz");
history.replace("/foo");
```

在设计完成四类导航接口后，由于调用最为频繁的页签内跳转使用 history.push("/baz")的方式，而在 9.14.1 节中设置了全局的页面缓存。在默认情况下调用 history.push("/baz")将缓存页面，与上述页签内跳转、页面的销毁方式有一定冲突，因而提供 history 的改写方法，将 history 的默认导航行为进行一定程度的改写：

```
export function historyEnhance(history: History) {
  const methods = ["push", "replace"];
  // 用以保存原始的导航方法
  const originHistoryMethods = {};
  methods.forEach(method => {
    originHistoryMethods[method] = history[method];
  });
  methods.forEach(method => {
    history[method] = (...args) => {
      // state 中将包含_option.unmountPath、_option.remountPath
      const { path, state, option } = getLocation(args, history);
      if (typeof option === "object") {
        if (option.forceRemount && option.preserveDom) {
          // 由于删除了 unmountPath, unmountPath 路径的 DOM 将保留
          // remount 路径由 getLocation 函数提供
          delete state._option.unmountPath;
        }
        if (!option.forceRemount && option.preserveDom) {
          // 在不设置 state 时，由于全局 standAlone，将保存 DOM
          state._option = {};
        }
      }
      originHistoryMethods[method](path, state);
    };
  });
  return history;
}
```

该方法读取导航参数的第三个参数，判断 forceRemount、preserveDom 属性，用以确定进行何种导航操作，在实现上需要更改 getLocation 返回的 state._option 的值。由于默认的 history.push("/a") 方法为第四类导航，因此需要通过 getLocation 与 getRefreshState 提供对应的 state，以满足组件的重新挂载与销毁：

```
import { History, parsePath, LocationDescriptor } from "history";
function getLocation(args: any[], history: History) {
  let realpath: string;
  let state: any;
  let option: any;
  if (typeof args[0] === "object") {
```

```
        const location = args[0];
        option = args[1] || {};
        realpath = history.createHref(location);
        state = getRefreshState(
            location.pathname,
            location.state,
            history.location
        );
    } else {
        realpath = args[0];
        option = args[2] || {};
        state = getRefreshState(
            parsePath(realpath).pathname,
            args[1],
            history.location
        );
    }
    return { path: realpath, state, option };
}
export function getRefreshState(path, state, location) {
    let stateCopy = state;
    if (!state) {
        // 提供默认 state
        stateCopy = {
            _option: {
                // 目标路径重新挂载
                remountPath: path,
                // 原路径销毁
                unmountPath: location.pathname
            }
        };
    } else {
        if (!state._option) {
            stateCopy._option = {
                // 目标路径重新挂载
                remountPath: path,
                // 原路径销毁
                unmountPath: location.pathname
```

```
      };
    }
  }
  return stateCopy;
}
```

与此同时，可基于上述 history，提供扩展的 Link 组件，以满足页面内组件导航的场景：

```
interface LinkPropsExt extends LinkProps {
  preserveDom?: boolean;
  forceRemount?: boolean;
}

export function LinkExt(props: LinkPropsExt) {
  const { match, history } = useContext(__RouterContext);
  return (
    <Link
      {...props}
      onClick={e => {
        const path = resolve(props.to as LocationDescriptor, match);
        e.preventDefault();
        const historyMethod = props.replace ? history.replace : history.push;
        historyMethod(
          path,
          typeof props.to === "object"
            ? {
                forceRemount: props.forceRemount,
                preserveDom: props.preserveDom
              }
            : null,
          {
            forceRemount: props.forceRemount,
            preserveDom: props.preserveDom
          }
        );
      }}
    />
  );
}
```

2. 命名导航

在使用 history 导航时,除了使用路径导航,如果为每个路由配置命名,如下所示:

```
const routeConfig = {
  path: "/user/info",
  component: UserInfo,
  // 命名配置
  name: "userInfo"
};
```

则可通过索引查询到此路由配置的地址信息。同理,也可以使用命名的方式进行导航:

```
// 导航到名为 userInfo 的路由路径
history.push({
  name:'userInfo'
})
```

若希望使用命名导航,则需要知晓完全的路由配置,只有在知道了整个路由配置树的情况之后,才可通过名称索引的方式获取到该命名配置所对应的路径。为此可实现通用的 walk 方法用于遍历配置树,拥有该方法,也可以在其他需要遍历配置树的场景中使用:

```
// 深度优先,遍历路由配置树
export function walk(
  // RouteExt 为路由配置 typescript 类型
  route: RouteExt,
  currentPath: string,
  parentRoute: RouteExt | null,
  cb: (r: RouteExt, currentPath: string, parentRoute: RouteExt | null) => void,
  predictor?: (eachRoute: RouteExt) => boolean
) {
  // 对每个配置调用回调,参数依次为当前路由配置、当前路径和父级路由配置
  cb(route, currentPath, parentRoute);
  // 此函数可用于结束递归遍历过程
  if (predictor && predictor(route)) {
    return;
  }
  // 对每个子路由递归调用 walk 函数
  route.routes &&
    route.routes.forEach(subRoute => {
      const routePath = `${processPath(currentPath)}${processPath(
```

```
      subRoute.path
    )}`;
    walk(subRoute, routePath || "/", route, cb, predictor);
  });
}
```

通过 walk 方法遍历路由配置树的方法，可从根路径开始，搜集路由配置树的路径索引信息：

```
export function getAllPaths(route: RouteExt) {
  // 路径索引 Map
  const pathNameMap = {};
  // 对配置树中的每个配置建立路径的索引
  walk(route, (route.path as string) || "/", null, (route, currentPath) => {
    pathNameMap[route.name || (route.path as string)] = currentPath;
  });
  return pathNameMap;
}
```

由此，可通过调用 getAllPaths(rootRoute)，搜集路由配置树所有的路径索引信息，并提升 history 的 push 与 replace 方法：

```
export function namedHistoryEnhance(history: History, route: RouteExt) {
  // 搜集所有命名导航
  const namedPathCache = getAllPaths(route);
  const methods = ["push", "replace"];
  // 用以保存原始的导航方法
  const originHistoryMethods = {};
  methods.forEach(method => {
    originHistoryMethods[method] = history[method];
  });
  methods.forEach(method => {
    history[method] = (...args) => {
      const path = args[0];
      if (typeof path === "object" && path.name) {
        const location = {
          ...path,
          // 索引到该命名配置所对应的路径
          pathname: namedPathCache[path.name]
        };
        args[0] = location;
```

```
        originHistoryMethods[method](...args);
      };
    });
  }
```

提升后的 history.push/replace 方法，如果第一个参数为对象，且 name 属性不为空，则会索引到该命名配置所对应的路径，并最终导航到此路径下。

9.14.3 使用页签组件

9.10 节曾介绍过页签机制与路由的结合，页签组件需要传入应用所有的页签项，本节将搜集所有的页签配置，并应用到页签组件中。在使用页签时，需要确认页签展示的名称、是否可关闭等信息。这类信息可通过配置进行维护，但由于已经存在一份路由配置，如果再为页签维护一份配置，则显得较为冗余。

考虑在原有路由配置上扩展属性，增加 tab 属性，该属性值为页签配置对象，类型为 TabConfig，接口如下：

```
import { RouteProps } from "react-router";
import { Location } from "history";
// 页签配置类型
export interface TabConfig {
  closable?: boolean;
  // 可配置页签的初始导航地址
  location?: Location;
  renderWithOutMatched?: boolean;
  // tab 上的名字
  tabName: string;
  tabIsMatch?: (currentLocation: Location) => boolean;
}
// 路由配置
export interface RouteConfig extends RouteProps {
  redirect?: string;
  standAlone?: string;
  // 子配置
  routes?: RouteConfig[];
  // 命名配置
  name?: string;
```

```
  isIndexComponent?: boolean;
  // 路由配置增加该路由的页签配置
  tab?: TabConfig;
}
```

在配置页签时允许配置页签展示文字（类似于 document.title）、是否可关闭、初始导航地址和额外的匹配函数等，相关内容可查看 9.10 节的页签配置。同时，在浏览器中，在一个页签下操作页面时，该页签处于激活状态。9.10 节提供了页签默认的激活判断函数，也可通过 tabIsMatch 自定义页签激活态判断逻辑，这些接口都由 TabConfig 定义，且附属于路由配置中的 tab 属性。

在路由配置中使用 tab 属性配置页签，好处是其天然地继承了路由配置的路径信息，并可使用路由配置路径作为页签的唯一标识符。如下示例，在路径/user 配置下有两个子配置，路径分别为/user/info、/user/setting，并分别配置了页签"用户中心""用户设置"：

```
const routeConfigs = {
  path: "/user",
  routes: [
    {
      path: "/info",
      // 用户中心页签，以/user/info 作为唯一标识符
      tab: {
        tabName: "用户中心"
      },
      routes: [
        {
          // 收藏列表
          path: "/favlist"
        },
        {
          // 个人空间
          path: "/space"
        }
      ]
    },
    {
      path: "/setting",
      // 用户设置页签，以/user/setting 作为唯一标识符
      tab: {
        tabName: "用户设置"
```

```
      },
      routes: [
        {
          // 标签设置
          path: "/tags"
        },
        {
          // 隐私设置
          path: "/private"
        }
      ]
    }
  ]
};
```

"用户中心"页签将管辖以 /user/info 作为路径开头的路径,如收藏列表 /user/info/favlist、个人空间 /user/info/space 页面都在"用户中心"页签下。同理,/user/setting/tags、/user/setting/private 页面都在"用户设置"页签下。

9.10 节的页签匹配函数默认为 startsWith:

```
// 规定了路径以 tabUrl 开头匹配页签
function defaultMatch(tabUrl: string, location: Location) {
  return location.pathname.startsWith(tabUrl);
}
export function tabMatch(tab, currentLocation) {
  if (tab.tabIsMatch) {
    return tab.tabIsMatch(currentLocation);
  }
  return defaultMatch(tab.tabUrl, currentLocation);
}
```

/user/info 没有传入额外的页签匹配函数,因此使用上述默认的页签匹配函数。由于 /user/info/space、/user/info/favlist 都以 /user/info 作为路径开头,因而它们都属于 /user/info 所配置的"用户中心"页签。

在上述示例中,页签匹配函数使用了默认的匹配函数。在通常情况下,在路由配置设计得当的情况下没有问题,一个父路由配置提供页签,其所有的子配置都在此页签的管理下;但是在某些情况下,更希望自定义页签的匹配命中函数,如根路径"/"对应了首页页面,同时也存在"首页"页签。若使用默认的页签激活态判断函数,则由于任意路径都以"/"开头,因此任意路径都会激活"首页"页签,这与期望不符。这时可传入自定义的页签匹配函数:

```
const routeConfig = {
  path: "/",
  component: AppLayout,
  routes: [
    {
      component: <Home />,
      //为父配置提供默认子组件
      isIndexComponent: true,
      tab: {
        tabName: "首页",
        // 首页页签常驻存在,不可关闭
        closable: false,
        // 初始导航路径
        location: { pathname: "/" },
        // 不命中路径也渲染页签,保证此页签初始渲染
        renderWithOutMatched: true,
        tabIsMatch: location => {
          // 仅在路径为根路径时激活此页签
          return location.pathname === "/";
        }
      }
    }
  ]
};
```

上例在配置中设置了 tabIsMatch 函数,在页签组件渲染页签并判断页签是否激活时,会使用传入的此函数进行页签激活态判断,这里不使用默认的页签命中函数,在仅判断路径完全等于"/"时,此页签才算匹配成功,进而页签才可激活。

"首页"页签除了传入 tabIsMatch 用于自定义页签命中态判断,还设置了 closable 为 false,表示页签不可关闭;设置了 location,表示初次单击页签时的导航路径;设置了 renderWithOutMatched 为 true,表示页签在渲染上应无条件渲染而不考虑路由命中情况。

9.10 节的页签组件,希望接收一个应用中包含所有页签信息的数组,以便在其判断渲染时使用。在上一节提供了 walk 方法用于遍历路由配置,可使用此方法收集页签信息:

```
export function collecteTabs(route: RouteExt, currentPath: string) {
  const tabs = [];
```

```
walk(
  route,
  currentPath,
  null,
  (eachRouteConfig, currentPath, parentRoute) => {
    if (eachRouteConfig.tab) {
      tabs.push({
        ...eachRouteConfig.tab,
        // 传入路由配置的路径作为页签的页签标记
        tabUrl: currentPath
      });
    }
  }
);
return tabs;
}
```

函数接收路由配置，且返回路由配置下的所有页签信息。通过引入此收集函数，结合 9.10 节的页签逻辑组件，组件可获得页签支持：

```
export default function AppLayout(props) {
  // 每次渲染都重新收集，兼容 route 变化场景
  const tabs = collecteTabs(props.route, props.routePath);
  return (
    <>
      {/* …… */}
      {/* TabsProvider 可查看 9.10 节 */}
      <TabsProvider allTabs={tabs || []}>
        {/* Tabs 仅负责 UI TabsProvider 提供页签逻辑 */}
        <Tabs>{props.children}</Tabs>
      </TabsProvider>
    </>
  );
}
```

接下来，考虑页签中的导航行为。在 9.10 节页签机制的介绍中，曾介绍过页签的两类导航行为，分别是关闭页签导航及切换页签导航；关闭页签导航又分为关闭当前激活页签和关闭非激活页签。按照页签设计，关闭页签时应销毁页签对应的 DOM 元素，而切换页签时应缓存页面，即保留页面 DOM 元素。结合 9.14.1 节中的销毁机制，以及 9.14.2 节中的页签切换导航，可在

TabsProvider 中引入如下导航：

```ts
// 传入 closeTab 函数给子组件，子组件在关闭页签时调用此函数
function closeTab(tab: Tab) {
  if (renderdTabs.length <= 1) return;
  const tabIndex = renderdTabs.indexOf(tab);
  if (tabIndex !== -1) {
    if (!tab.isActive) {
      // 关闭的 tabpane 为非激活的 tabPane
      renderdTabs.splice(tabIndex, 1);
      // 使用 history.replace 替换，仅为了触发一次路由的更新进而销毁 DOM 元素，
      // 页面地址及历史栈无变化
      history.replace(history.createHref(currentLocation), {
        _option: {
          // 销毁当前页签路径所对应的组件
          unmountPath: tab.location.pathname
        }
      });
    } else {
      // 关闭的 tabpane 为当前激活的，需要找到距离关闭项最近的 tabpane 进行激活
      const nearestTab = findNearestTab(tab, renderdTabs);
      renderdTabs.splice(tabIndex, 1);
      history.push(history.createHref(nearestTab.location), {
        _option: {
          // 销毁当前页签路径所对应的组件
          unmountPath: tab.location.pathname
        }
      });
    }
  }
}
// 传入 activeTab 函数给子组件，子组件在切换页签跳转时可调用此函数
function activeTab(tab: Tab) {
  // 保留 DOM 元素，不重新渲染
  history.push(history.createHref(tab.location), null, {
    forceRemount: false,
    preserveDom: true
  });
}
```

在页签间切换时调用 activeTab 函数，该函数使用 history.push，不强制刷新页面，forceRemount 为 false，同时设置 preserveDom 为 true，确保 DOM 得到保留。同时，在关闭页签时，state._option.unmountPath 为即将关闭页签中保存的 location 地址，这样保证了在关闭页签时，该地址下的组件能得以销毁。

在丰富了页签逻辑组件的内容后，tabs 页签 UI 组件仅需提供页签的 UI 实现并使用 TabsProvider 提供的接口即可：

```
export default function(props) {
  if (!props.tabProp || !props.tabProp.tabs) return null;
  return (
    <div className="tab">
      <div>页签区域</div>
      <div className="biaoqian">
        {props.tabProp.tabs.map((tab, index) => {
          return (
            <TabItemHeader
              key={index}
              active={tab.isActive}
              onJump={() => {
                props.tabProp.activeTab(tab);
              }}
              onClose={() => {
                props.tabProp.closeTab(tab);
              }}
              closable={tab.closable}
              title={tab.tabName}
            />
          );
        })}
      </div>
      {props.children}
    </div>
  );
}
```

9.14.4 页签栈维护

1. 维护页签栈背景

在子项目独立开发过程中，在子项目内能使用 history.goBack 进行回退，但是在页签内与主项目联合时使用 history.goBack 回退将与预期不符，原因在于主子项目联合使用时，浏览器历史栈的变化更加复杂，下面通过示例进行说明，如图 9-9 所示。

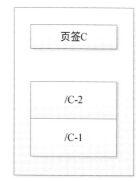

图 9-9　各页签维护路径

在图 9-9 中，页签 A 维护了/A-1、/A-2 路径，页签 B 维护了/B-1、/B-2 路径，页签 C 维护了/C-1、/C-2 路径。

若有以下导航路径：使用 push 导航到/A-1，再导航到/A-2，留在 A 页签内；再使用 push 导航到/B-1，跳转到 B 页签，并在 B 页签内操作导航到/B-2 后，又单击 A 页签切换到了页签 A 的/A-2 路径，浏览器导航产生的历史栈如图 9-10 所示。

对于 A 页签的/A-2 路径对应的页面，如果在/A-2 页面不加处理就调用 history.goBack()，将会回到 B 页签的/B-2 页面，如图 9-11 所示。

这与/A-2 页面期望的行为不符，期望的/A-2 页面的 history.goBack 调用应该回到/A-1 页面。

为了解决这类问题，考虑两种解决思路：

第一种解决思路是在页签 B 切换到页签 A 的过程中，使

图 9-10　浏览器导航产生的历史栈

栈指针重新回到真正的/A-2 位置，而不是新增加或替换历史栈记录，如图 9-12 所示。

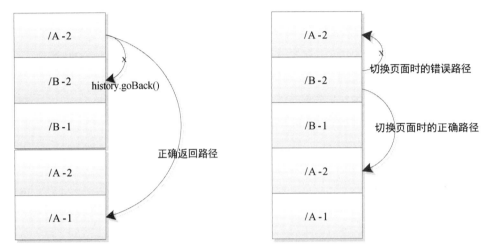

图 9-11 history.goBack()错误返回　　　　图 9-12 页签切换时变更栈指针

要使栈指针回到/A-2 页面，可以通过维护一个整个应用的历史栈，判断应该移动的 delta 距离，调用 go 移动栈指针进行实现，在 history 源码中有使用到此方法。但这样做存在一个问题，回到/A-2 页面后，虽然调用 history.goBack()能使得页面回到/A-1，但是如果此时再次调用 push 方法添加某个历史栈记录，如跳转到/C-1 页面，则会使得历史栈丢失 B 页签的导航历史记录，如图 9-13 所示。

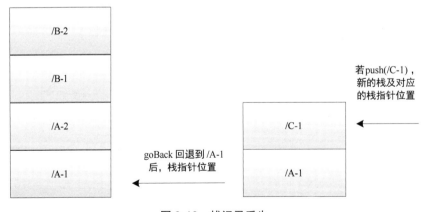

图 9-13 栈记录丢失

这样丢失了 B 页签下的导航信息，也将无法找回了。

第二种解决思路是忽略全局历史栈，为每个页签维护一个独立的历史栈，模拟浏览器原生页签的实现，每个原生页签都维护了自身的页签历史栈。在切换历史栈的过程中，每个页签对应的历史栈不会有变化。在 9.14.2 节中，曾介绍过四类导航，这里分析一下各类导航对历史栈的影响。首先分清一下导航实现：

- 菜单导航，这类导航类似于在浏览器中新开原生页签（菜单对应页签存在），或者切换到某个已存在的原生页签上（菜单对应页签存在），对应于浏览器导航新开原生页签产生一个空页面或者某些浏览器插件会初始化一个历史记录。总结起来，这类导航仅对初始化历史栈有影响。
- 页签间切换导航，类似在浏览器中切换各个原生页签，各个原生页签的历史栈均不受影响。
- 关闭页签导航，类似于在浏览器中关闭原生页签，对于一些可恢复关闭原生页签的浏览器，其原生页签的历史栈会保留，但页签系统在设计上没有设计关闭页签恢复机制，这里会将页签的历史栈清除。
- 页签内页面间导航，在浏览器中，各原生页签拥有自身的历史栈，历史栈的变化都在各自的页签内，因此本节页签系统的自身历史栈也仅维护本页签内的跳转页面，这类跳转在业务中使用最多，最为频繁。

在梳理完各导航行为后，栈的维护也变得清晰，由于为每个页签维护了历史栈，也带来了两个好处：

一是可以从页签历史栈获得用户在单个页签下的所有导航记录。若当前的浏览器栈为/A-1、/A-2、/B-1、/B-2，栈指针指向/B-2，A 页签的历史栈为/A-1、/A-2，B 页签的历史栈为/B-1、/B-2。由于知道了 A 页签的历史栈为/A-1、/A-2，则若从/B-2 导航到/A-2 后，虽然浏览器维护的栈/A-1 在栈底，用传统的 goBack 方法无法回退到/A-1；但在/A-2 页面，可通过自身页签维护的栈知道 goBack 应该回到/A-1 页面，通过拦截 goBack 方法，将 replace 替换为/A-1 可以做到这一点。

二是此方案还能解决浏览器历史栈被清理的问题。假设现有的历史栈为/A-1、/A-2、/B-1、/B-2，则 A 页签的历史栈为/A-1、/A-2，B 页签的历史栈为/B-1、/B-2，栈指针指向/B-2。若通过移动栈指针 go(-3)使得栈指针移动到栈底，回到了页签 A 下的/A-1 页面，此时如果调用 push('/C-1')打开 C 页签的/C-1 页面，由于浏览栈的栈机制，会使得指针移动一位，同时使得/C-1 作为新的栈顶，这样将会导致浏览器中栈记录/B-1、/B-2 的丢失。如果不自行维护栈记录，假若单击页签 B 回到了/B-2 页面，这时从浏览器栈中也将无从获得/B-2 的上一个页面为/B-1，因为/B-1 已经被清理了。但如果有栈记录，则在/B-2 页面，还可从栈记录中获取到 goBack 正确的返回地址应为

/B-1。如图 9-14 所示,各页签维护了自身的历史栈。

图 9-14　各页签栈记录

但此种方案也存在一个缺点,即浏览器自身历史栈后退行为,对于大多数在页签内操作的场景,虽然浏览器的"前进"和"后退"都符合预期,但在切换页签后,对浏览器栈来说,如果切换页签的行为为 push,则其栈的记录数量不会减少,浏览器栈从头到尾记录了用户所有的操作,栈的记录呈现分段式结构,如图 9-15 所示。

在切换页签后,如当前位于栈顶/B-2,单击浏览器的原生返回按钮将回到/C-2,与在/B-2 页面内调用 history.goBack 回到/B-1 的期望有出入。这需要注意,需要评估这类问题的影响。

同时,如果各页签的历史栈存在内存中,则存在刷新后历史栈丢失问题。针对此类丢失问题,可以通过持久化存储历史栈来解决,这样即便刷新后,用户产生的历史栈也不会丢失。本节示例选用 localstorage 作为持久化存储的位置。

2. 页签栈实现

在实现上秉承 go 不改变历史栈,只对能改变历史栈记录的 push、replace 方法进行处理。

为了获得拦截 goBack 方法的能力而又不影响 history 的接口定义,需要对 history 有一定程度的修改,

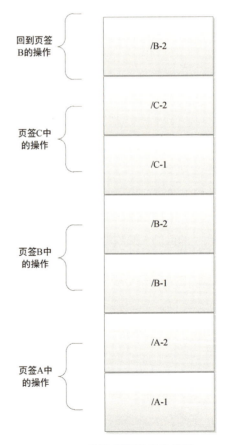

图 9-15　浏览器原生栈的分段

在 9.8 节中曾介绍过为 history 引入前置中间件，如：

```
// 调用 next 表明进行下一个中间件，与 Redux 类似
applyMiddleware(history, [middware1]);
export async function middware1(next, state) {
  console.log("第一个中间件");
// 如果希望中间件拦截 history 等方法，则不用调用 next
  await next();
}
```

如果在中间件中不调用 next 方法，则真实的跳转将不会执行，同时如果 next 方法第一个参数传递为字符串，则会重定向到该地址，因此可以为 history 引入中间件解决 goBack 拦截问题。与此同时，也可在中间件中获得 push 等导航的地址，进而维护各自页签的栈记录。

在中间件中使用栈，希望使用栈的时候接口尽量简化，应该仅需调用 push、replace 等方法就可完成栈的维护，因此设计 Stack 类，用于维护栈的相关逻辑：

```
class Stack {
  constructor(stackIdentify, stack) {
    this.stackIdentify = stackIdentify;
    if (stack) {
      this.stack = stack;
    }
  }
  stackIdentify;
  stack = [];
  initStackIfNeeded(stackItem: string) {
     // 初始化
    if (!this.stack.length) {
      this.stack.push(stackItem);
    }
  }
  push(stackItem: string, currentPath: string) {
    const prevIndex = this.stack.lastIndexOf(currentPath);
    // 更新栈，舍弃栈中当前记录的后续
    const nextPaths = this.stack.slice(0, prevIndex + 1);
    // 入栈
    nextPaths.push(stackItem);
    this.stack = nextPaths;
    return this.stack;
```

```
    }
    replace(stackItem: string, currentPath: string) {
      if (!this.stack.length) {
        this.stack.push(stackItem);
      }
      const prevIndex = this.stack.indexOf(currentPath);
      // 替换栈记录
      if (prevIndex !== -1) this.stack[prevIndex] = stackItem;
      return this.stack;
    }
    getStackItemBy(relativeStackItem, delta) {
      if (this.stack.length === 0) {
        return;
      }
      // 获取某个栈记录的相对偏移
      const prevIndex = this.stack.lastIndexOf(relativeStackItem);
      return this.stack[prevIndex + delta];
    }
    clearStack() {
      this.stack = [];
    }
}
```

栈在内容上使用 string 标志每一个栈的记录，这个栈元素可以为随机 key 值（如 browserHistory 所维护的），或者栈元素也可以为 hashHistory 中的路径信息（如 path）。对于 Stack 类，其不应该关心标识所代表的意义，设置为 string 即可（为了便于理解，本例使用路径 path 作为栈内的元素项）。在页签中切换，或者单击菜单激活页签等场景，栈的内容应该不变，仅在页签内跳转时才维护栈的记录，因此需要区分各种跳转场景。对于 Stack 类来说，如果某一个页签未被激活打开，其栈内容为空，当单击菜单激活该页签时，栈也应初始化，栈的初始记录应为当前激活的页签对应的地址，而当再次单击已激活页签时，栈的内容应该不变，则 Stack 类还需要一个 init 栈的方法。由于在 goBack 时，需要获知页签栈中当前地址的上一个栈记录，作为真正的回退地址，Stack 类还需要一个获取相对地址的方法，如 getStackItemBy。该方法传入一个已存在的栈记录，以及一个相对距离，可得到栈中对应偏移距离的另一个栈记录。

Stack 类中 push 与 replace 方法都模拟了浏览器的实现，会获取到当前的地址信息，推断其在栈中的位置。如果是 push，则入栈新的地址并更新栈顶；如果是 replace，则替换原有栈中位置的地址信息。

注意，replace 方法中也有入栈记录：

```
if (!this.stack.length) {
  this.stack.push(stackItem);
}
```

作为特殊情况，如果以 Redirect 或 replace 的方式初次打开页签，且没有走到初始化 initStackIfNeeded 逻辑，则需要在 replace 中初始化栈记录，这是特殊的初始化逻辑。

在 9.10 节页签系统中曾介绍过每个页签都有一个唯一的标识，页签以该标识作为唯一的凭证，而页签的栈记录也可使用页签标识作为索引。为了统一管理所有的页签栈，引入 StackManager 类：

```
class StackManager {
  constructor() {
    window.addEventListener("beforeunload", () => {
      // 在离开页面前保存所有打开页签的栈信息
      // 由于保存到了 localstorage 中，也可以支持所有页签栈的恢复，以及渲染页签列
      // 表的恢复
      localStorage.setItem("_stack", JSON.stringify(this.stackCache));
    });
  }
  stackCache = {};
  getStack(stackIdentify: string) {
    if (!this.stackCache[stackIdentify]) {
      const stack = safeLocalStorageGet("_stack");
      // 实例化 Stack
      this.stackCache[stackIdentify] = new Stack(
        stackIdentify,
        // 初始化 Stack 时从 localstorage 中获取对应的栈数据进行初始化
        _.get(stack && stack[stackIdentify], "stack") || []
      );
    }
    return this.stackCache[stackIdentify];
  }
}
```

StackManager 将以 stackIdentify 作为索引保存所有的栈信息到 stackCache 中，使用方可通过 getStack 获取到栈的单例实例。在 StackManager 中，每个栈实例仅实例化一次，之后都由缓存提供，保证每个页签栈仅有一个实例。

在设计好栈的管理者后，可引入 history 中间件（9.8.2 节曾介绍过中间件定义），获知所有的页面导航情况，并可在中间件中维护各个页签栈，如入栈、出栈等操作；同时，也拦截页签导航 goBack，在拦截后从页签栈中获得真实的后退地址，进行重定向。

中间件的执行流程为：
- 根据导航过程中的信息获取到要处理的页签栈。
- 处理对应的页签栈，如更新页签栈内容。
- 如果为 goBack 方法，则拦截并重定向，重定向地址根据当前页签栈与当前的地址决定。

要处理的页签栈有两种情况：如果为关闭页签，则要处理的页签栈为被关闭的页签的历史栈；如果要处理的页签栈为 "to"，即要导航到的地址所对应的页签栈，由于中间件提供了 to，可根据导航的目标 "to"，获取到 "to" 所对应的页签标识，进而获取到该页签的 stack 实例，所有的栈操作均在此实例上进行。stackHistoryMiddware 中间件如下：

```
// collecteTabs 在 9.14.3 节有过介绍，获取路由配置中定义的所有页签信息
const tabs = collecteTabs(routesConfig[0], "/");
export async function stackHistoryMiddware(next, { history, method, from, to, args }) {
    const closePath = _.get(args[1], "_option.unmountPath");
    // 关闭页签时应该以被关闭页签的路径作为目标路径，因为要处理的是被关闭的页签的栈
    const destTo = closePath || to;
    // createLocation 可解决相对路径问题，如 to 为 "../one" 时返回的路径为绝对路径
    const destLocation = createLocation(destTo, null, null, history.location);
    // tabMatch 从页签部分导入，用于判断某个页签是否命中，曾在 9.14.3 节有过介绍
    // 根据要处理的地址获取要处理的页签
    const destTab = tabs.find(tab => tabMatch(tab, destLocation));

    if (destTab) {
      // destTab.tabUrl 为页签的唯一标识
      const stack = stackManager.getStack(destTab.tabUrl);
      const href = history.createHref(destLocation);
      if (args[2] && !closePath) {
        // 非页面内跳转导航，如页签间的切换、单击某菜单激活页签等
        stack.initStackIfNeeded(href);
        await next();
        return;
      }
      if (closePath) {
```

```
      // 关闭页签时产生的导航
      stack.clearStack();
      await next();
      return;
    }
    if (method === "push" || method === "replace") {
      // 页签内跳转导航
      stack[method](href, history.createHref(from));
      await next();
      return;
    }
    if (method === "goBack") {
      // 获得真实的回退地址
      const goBackHref = stack.getStackItemBy(
        history.createHref(from),
        // 相对距离，delta 为-1
        -1
      );
      // 重定向到真实地址
      await next(goBackHref);
      return;
    }
  }
  await next(null, true);
}
```

在清楚中间件的执行流程后，还需要厘清何时进行栈的维护。在上一节中，曾介绍过页面内的四种导航，分别是菜单导航、页签间切换导航、关闭页签导航和页签内页面间导航，它们在调用参数上存在一定区别。对于菜单导航与页签间切换导航，由于有重新挂载页面与保存 DOM 的需求，在 history 调用时增加了额外的参数：

```
// 菜单导航
history.push(path, state, {
  // 强制跳转 path 重新渲染
  forceRemount: true,
  // 保留其他 DOM
  preserveDom: true,
});
// 页签间切换导航
```

```
history.push(path, state, {
  // 不强制重新挂载
  forceRemount: false,
  // 所有的 DOM 应该保留
  preserveDom: true,
});
```

而关闭页签导航指定了需要销毁的路径：

```
// 关闭页签导航
history.push(history.createHref(nearestTab.location), {
  _option: {
    unmountPath: tab.location.pathname
  }
});
```

仅页签内页面间导航无须特殊参数：

```
// 页签内页面间导航
history.push("/foo");
```

在上述导航中，菜单导航与页签间切换导航仅在页签栈为空的情况下，作为页签的首次导航，需要对页签栈产生影响，因此需对栈做初始化处理，入栈一次。在初始化栈后，不需要更改栈记录。因此调用了 stack.initStackIfNeeded 方法，仅在栈为空时添加栈记录。可根据 history 参数列表中第三个参数的有无来判断是否属于这两类导航。

而关闭页签导航在 state 中存在 unmountPath 参数，遇到这类导航，需要清除被关闭的页签栈。

真正产生页签栈频繁变更的导航是页签内页面间导航，此时可调用 stack 的 push 或者 replace 方法，维护对应页签栈的栈记录。

在拦截 goBack 的实现上，通过中间件参数中 method 的判断，如果为 goBack 调用，则可对其进行拦截。拦截后需要进行重新跳转，在跳转地址的选取上，根据当前路径获取其栈中上一个地址，即真正的回退地址 goBackHref，使用 next(goBackHref)进行重定向，达到调用 goBack 方法也能正确返回的目的。

最后，为了使得栈记录可以被恢复，使用 localstorage 进行页签栈的保存。在保存时机上，每次关闭应用时将页签的栈情况快照下来，保存最后关闭应用前的页签栈情况到 localstorage 中。由于使用 localstorage 保存了栈的记录信息，所以就算刷新页面后，也不仅可以从 localstorage 中获知 goBack 回退时应真正回退到的页面地址，所有打开过的页签也都可以从 localstorage 中获取。

有了这个数据，即便关闭应用，再次进入应用时也可将应用关闭前的所有打开页签进行恢复。同样持久化页签栈使得用户即便强制刷新页面，页签栈依旧可以恢复。这与初始加载类似，刷新页面没有主动的导航发生，中间件方法没有被调用，若原先导航产生了页签栈，虽然栈的内容由于保存在内存中被用户刷新页面后清空，但由于在页面刷新前 StackManager 对栈内容进行了持久化，保存到了 localstorage 中，因此 stack 重新实例化时会从 localstorage 中恢复栈内容：

```
getStack(stackIdentify: string) {
  // 内存已被清空
  if (!this.stackCache[stackIdentify]) {
    // 获取持久化信息
    const stack = safeLocalStorageGet("_stack");
    this.stackCache[stackIdentify] = new Stack(
      stackIdentify,
      // 从 localstorage 中恢复栈内容
      _.get(stack && stack[stackIdentify], "stack") || []
    );
  }
  return this.stackCache[stackIdentify];
}
```

9.15 小结

本章结合 React Router 的基础知识，介绍了多个具备实用性与可扩展性的示例，包括：组件层面的滚动恢复，路由与组件的结合案例；导航层面的异步导航，导航中间件；高阶组件的路由能力提升，如路由生命周期、路由页签机制；状态管理与路由的结合，以及可在小型项目中使用的简化路由，如自定义路由 Hook，组件路由化；中大型项目可使用的微服务路由、配置化路由；中大型项目的配置化路由实战。本章通过上述各实战示例的引入，为读者学习与理解 React Router 提供了全面丰富的素材。

通过本章系统性的学习，读者可全方位地学习与理解前端路由，提升对前端路由的整体认识，掌握前端领域路由的设计思路与方法，在深入理解前端路由设计的同时，为今后的学习及工作打下牢固的基础。

参考文献

[1] https://zh.wikipedia.org/wiki/%E9%9D%A2%E5%8C%85%E5%B1%91%E5%AF%BC%E8%88%AA.

[2] https://github.com/supasate/connected-react-router.

[3] https://github.com/alisd23/mobx-react-router.

附录 A
从 React Router v3.x 迁移到 React Router v4.x 及以上版本

1. 去中心化

在 React Router v3.x 中，仅有一个 Router 组件，需要提供 history 对象作为 Router 的属性，所有的 Route 都需要写在 Router 的 routes 属性中或者作为 Router 的子组件以作为中心化的配置。在 React Router v4.x 及之后的版本中，可以有不同的 Router，如 BrowserRouter、HashRouter、MemoryRouter，且没有集中的路由配置：

```
<BrowserRouter>
  <div>
     {/* 任何位置都可使用 Route */}
    <Route path='/about' component={About} />
     <div>
    <Route path='/contact' component={Contact} />
 </div>
  </div>
</BrowserRouter>
```

2. 去配置

在 React Router v3.x 中，Route 并不是一个组件；相反，程序中所有的 Route 元素仅用于创建路由配置对象，并且将 Route 转化为对应的路由配置对象存储在内存中：

```
/// React Router v3.x 中的 Route
<Route path='contact' component={Contact} />
// 相当于
```

```
{
  path: 'contact',
  component: Contact
}
```

使用 React Router v4.x，可以像常规的 React 应用一样使用路由组件，可以在任何需要渲染路由内容的地方渲染一个 Route 组件。Route 不再与 Router 有配置关系。仅当 Route 的 path 与当前的路径匹配时，它将会渲染 component、render 或 children 属性中的内容。

3. 去嵌套路由

由于 React Router 去中心化，同时也舍去了嵌套路由。在 React Router v3.x 中，Route 组件可互相嵌套：

```
<Route path='parent' component={Parent}>
  <Route path='child' component={Child} />
  <Route path='other' component={Other} />
</Route>
```

当嵌套的 Route 匹配时，React Router v3.x 将会使用子 Route 和父 Route 的 component 属性去渲染。而使用 React Router v4.x，子 Route 应该由父 Route 中的组件控制：

```
<Route path='parent' component={Parent} />
const Parent = () => (
  <div>
    <Route path='child' component={Child} />
    <Route path='other' component={Other} />
  </div>
)
```

4. 去生命周期 Hook

React Router v3.x 提供 onEnter、onUpdate 等路由生命周期方法。这些方法与 React 组件的生命周期方法有一些重合。

使用 React Router v4.x，通过 Route 渲染的组件，可以使用 componentDidMount 代替 onEnter，同样可以使用 componentDidUpdate 代替 onUpdate，可以使用 componentWillUnmount 代替 onLeave。

5. url params 变化

在 React Router v3.x 中,命名参数 params 的声明需要用圆括号包裹:

```
path="/entity/:entityId(/:parentId)"
```

而到了 React Router v4.x,params 的声明去掉了圆括号,且可使用正则表达式进行修饰:

```
path="/entity/:entityId/:parentId?"
```

同时,在 React Router v3.x 中,通过 props.params 可获取到命名参数匹配情况。而到了 React Router v4.x,props.params 将无法获取到匹配对象,需要使用 props.match.params,也要注意 props.match 为 null 的情况。

6. 渲染单个命中组件改为使用 Switch

在 React Router v3.x 中,可以在同一级指定多个子路由 Route,并且只会渲染匹配到的第一个 Route:

```
// React Router v3.x
<Route path='/' component={App}>
  <IndexRoute component={Home} />
  <Route path='about' component={About} />
  <Route path='contact' component={Contact} />
</Route>
```

React Router v4.x 通过 Switch 组件提供了相似的功能,当 Switch 被渲染时,它仅会渲染与当前路径匹配的第一个子 Route:

```
// React Router v4.x
const App = () => (
  <Switch>
    <Route exact path='/' component={Home} />
    <Route path='/about' component={About} />
    <Route path='/contact' component={Contact} />
  </Switch>
)
```

7. Redirect 变化

React Router v3.x 可使用 IndexRedirect:

```
// React Router v3.x
<Route path="/" component={App}>
  <IndexRedirect to="/welcome" />
</Route>
```

React Router v4.x 去掉了 IndexRedirect，如果希望进行类似跳转，则需要使用 Redirect 组件：

```
// React Router v4.x
<Switch>
  <Route exact path="/" render={() => <Redirect to="/welcome" />} />
  <Route path="/welcome" component={Welcome} />
</Switch>
```

在 React Router v3.x 中，Redirect 的重定向保存了 query 字符串：

```
// React Router v3.x
<Redirect from="/" to="/welcome" />
// /?source=google → /welcome?source=google
```

到了 React Router v4.x，如果希望在重定向时保存 query 字符串，则需要人工传入：

```
// React Router v4.x
<Redirect from="/" to="/welcome" />
// /?source=google → /welcome
// location 中存有 query
<Redirect from="/" to={{ ...location, pathname: "/welcome" }} />
// /?source=google → /welcome?source=google
```

8. Link 变化

React Router v3.x 的 Link 的 to 属性可为 null：

```
// React Router v3.x
<Link to={disabled ? null : `/item/${id}`} className="item">
  // item content
</Link>
```

在 React Router v4.x 及以后的版本中，to 属性不能忽略。如果希望实现与 React Router v3.x 一致的 Link，则可封装组件：

```
// React Router v4.x
import { Link } from "react-router-dom"
const LinkWrapper = props => {
  const Component = props.to ? Link : "a"
```

```
    return <Component {...props}>{props.children}</Component>
}
<LinkWrapper to={disabled ? null : `/item/${id}`} className="item">
  // item content
</LinkWrapper>
```

9. 帮助函数变化

React Router v4.x 将路径的匹配交由 path-to-regexp 库实现。React Router v3.x 中的 matchPattern、formatPattern 将失效。对于 matchPattern，可使用 matchPath 进行代替。formatPattern 通常用于将参数代入参数化路径，在 React Router v4.x 中可以通过使用 path-to-regexp 的 compile 函数实现：

```
// React Router v4.x
import pathToRegexp from "path-to-regexp"
const THING_PATH = "/thing/:id"
const thingPath = pathToRegexp.compile(THING_PATH);
<Link to={thingPath({ id: 1 })}>A thing</Link>
```

这需要引入 path-to-regexp 路径匹配库。

10. 版本变动小结

React Router v3.x 关注了 React 组件的创建、组合等，如 Router 可自定义组件的 createElement，applyRouterMiddleware 中间件关注了组件的嵌套组装，Route 提供了组件的异步加载及命名组件等，在组件上关注较多。在 React Router v4.x 之后，React Router 减少了对组件部分的关注度，将组件的加载、创建、嵌套等都交给了开发者。

React Router v3.x 为静态路由，其 Route 组件仅作为配置，路由在初始化后在内存中不再变动。而到了 React Router v4.x，与配置相关的 childRoutes、getChildRoutes、getIndexRoute、indexRoute 等都被移除，React Router v4.x 全为动态化路由。下面列出了 React Router v4.x 及以后版本中的几个变化点。

1）Router 变化

Router 不再接收静态配置 routes，除去了 createElement 方法传入，该方法可自行定义组件的创建，React Router v4.x 移除 createElement 方法后，由开发者自行定义组件。移除了 Router 中的 onUpdate、onError，在 React Router v4.x 及之后的版本中处理都可在组件中进行。移除了 Router 中的自定义渲染 render。

与 React Router v3.x 中的 Router 相比，React Router v4.x 及以后版本中的 Router 不负责收集路由配置，其仅负责接收 location 变化，并通知各 Route。

2）注入的 props 变化

Router 将不再注入组件中，原 Router 由 history 与 routerLeaveHook 等组成，原上下文中的 Router 的导航能力如 props.router.push 移动到 history 中，被 history 取代，如 props.router.push 变化为 props.history.push。同时 Router 中的跳转确认 setRouteLeaveHook 由 history.block 所取代，如跳转确认：

```
router.setRouteLeaveHook(this.props.route,()=>'确认跳转？')
// 将改为
history.block('确认跳转？')
```

在参数获取上，React Router v3.x 中 location 中的 query 与 action 被移除，query 需要自行解析，action 移动到了 history 中。params 及 routeParams 命名参数由 match 提供，不再作为 props 的一级属性。且当命中的 route 信息不再注入 props。

在路由命中数组上，当前命中的父子路径 routes 数组不再注入 props。在 React Router v3.x 中，当前命中的父子路径 routes 可方便地实现面包屑，被移除后需自行分段路由路径实现面包屑。

React Router v4.x 不再提供命名组件的注入。

3）Route 变化

Route 将真实渲染组件，不再仅是声明式，且 Route 的各属性进行了简化，生命周期 onEnter、onChange、onLeave 移除，React Router v4.x 的 Route 的生命周期由 React 组件提供。Route 除去了 components 命名组件，组件的组装交给开发者，除去了异步加载组件方法 getComponent、getComponents，React Router v4.x 之后路由异步加载页面功能可由高阶组件等实现。

4）history 变化

listenBefore 改变为 history.block，相对地也移除了 listenBefore 的异步串联。history 移除了 enhancer，如 useBasename 提供的 basename 支持、useBeforeUnload 提供的页面离开监听支持和 useQueries 提供的 query 解析能力等。enhancer 在 React Router v4.x 之后可自行实现，同时 getCurrentLocation 方法提供的 location 可由 history.location 提供。